T0142816

Excel-VBA

Tariq Muneer • Stoyanka Ivanova

Excel-VBA

From Solving Mathematical Puzzles to Analysing Complex Engineering Problems

Tariq Muneer🆔
School of Engineering and Built
Environment
Edinburgh Napier University
Edinburgh, Midlothian, UK

Stoyanka Ivanova🆔
Department of Computer-Aided Engineering
University of Architecture, Civil
Engineering and Geodesy
Sofia, Bulgaria

ISBN 978-3-030-97800-6   ISBN 978-3-030-94085-0   (eBook)
https://doi.org/10.1007/978-3-030-94085-0

This Springer imprint is published by the registered company Springer Nature Switzerland AG
The registered company address is: Gewerbestrasse 11, 6330 Cham, Switzerland

آنکھ جو کچھ دیکھتی ہے، لب پہ آسکتا نہیں
ہو حیرت ہوں کہ دُنیا، کیا سے کیا ہوجائے گی

Urdu verse by M Iqbal (1877–1938)

Lips can't utter, the vision that eyes grasp
Amazed I am of world's confounding
progression

Translation by Tariq Muneer

Acknowledgments

The authors would like to thank their publishers: Anthony Doyle and Sukumar Balaganesh for the help they have extended throughout the publication process.

T. Muneer would like to thank a number of people who have encouraged him to complete this project as they see a lot of benefits coming out of it. Some of those names that come to mind are (in alphabetical order): Afzal, Annu, Farhatullah, Fouad, Gazali, Hyder, Jamil, Khalid, Khursheed, Mobeen, Samad, Shajee, Taqe, Waheed.

Special thanks to Zafar Akber and Gulam Mohammed for providing valuable advice. Their kind words have helped the author throughout the pandemic and its demoralizing effects.

S. Ivanova would like to thank Plamen Chobanov, Evelina Ivanova, Vladimir Yakov, and Fantina Rangelova for providing support, valuable advices, and ideas for her chapters in the book.

Review of an Earlier Book by Prof. T. Muneer

Publication: Heat Transfer: A Problem Solving Approach, Tariq Muneer, Jorge Kubie, Thomas Grassie, Spon Press, London (Taylor & Francis), 2003, 388 pages, ISBN: 0-415-24109-3, $90. Includes a CD of example Excel problem solutions, and 101 line figures, 40 tables and 4 pages of color photographs.

Review published in: Heat Transfer Engineering, 28(6):598–600, 2006, Taylor and Francis Group
ISSN: 0145-7632 print/1521-0537 online, DOI: 10.1080/01457630701274367

Reviewed by: Professor Ralph L. Webb, Penn State University, University Park, PA 16802, United States

Review: This is a very interesting undergraduate text and definitely outside the mainstream "me too" books on the subject. What is unique about this book is the use of Excel to provide a comprehensive approach to problem solving. It includes a series of over 75 computer workbooks on an accompanying CD. Readers of this review may download several Excel example workbooks at http://www.soe.napier. ac.uk/heatxfer/heatxfer2.html. The CD contains two key work books:

1. Fluid property tables for 57 fluids. This 'Allprops' software is provided, courtesy of the University of Idaho.
2. Excel spreadsheets of the accompanying example problem solutions.

Although it may be common for students to use basic Excel procedures in solution of homework problems, this book offers the student a major step forward in using relatively complex Excel features. Examples of this include:

1. Electronic lookup tables for transport and thermodynamic properties.
2. Step-wise computations with dynamically linked graphs that enable numerical experimentation.
3. Matrix operations.
4. Solution of nonlinear system of equations experienced in multi-mode heat transfer.
5. Solution to linear and nonlinear optimization problems.
6. Finite element approach to obtain radiation view factors.

7. Use of in-built Visual Basic for Applications (VBA) language for handling complex iterative tasks. Also, use of Visual Basic environment for raster representation of cylindrical and spherical systems.
8. Use of Graphical User Interface (GUI) for solving one- and two-dimensional (Cartesian) conduction problems, producing temperature raster plots and exploring accuracy dependence on mesh size.

Chapter 2 describes the numerical procedures used in the Excel spreadsheets. This chapter provides important understanding of the advanced Excel subroutines used in the problem solutions. Each chapter of the book provides the basic equations and correlations applicable to the subjects addressed. Each chapter is accompanied by 5-to-7 homework problems. The student may revise existing Excel spreadsheets to solve these problems. The figures in the book do not compare to the "slick" figures used in major competing textbooks. Unfortunately, the book does not provide separate chapters on heat-exchanger design, boiling, and condensation. However, the chapter on "multi-mode heat transfer" is well done and linked to Excel examples on solving such problems.

I believe that the use of advanced Excel features to solve the examples and homework problems will introduce the student to a more comprehensive and up-to-date understanding of solutions to relatively complex solution methods. If the students use these tools, they will have a much more comprehensive understanding of state-of-the-art problem-solving methodology than gained by use of "hand solution" methods. The authors have done an outstanding job in this area. Hopefully, we will see more of such methodology in the major competing textbooks.

The authors, Tariq Muneer, Jorge Kubie, and Thomas Grassie are all located at Napier University, Edinburgh, and perform research in "energy engineering".

This unique book presents a practical textbook, written for both undergraduates and professionals. It presents its contents and solution methodology in a manner that is directly useful to a designer, while a researcher can use it to select Excel approaches to predictive models. The book does a very good job of meeting its objectives and is recommended to students, or others who desire state-of-the-art approach to problem solving. An instructor should find that the problem-solving methodology will enhance the depth of a student's learning experience. Future editions of the book should seek to expand the breadth of coverage to, at least, include heat-exchanger design. Inclusion of boiling and condensation would make the book more useful as a reference.

Contents

About the Authors

Prof. Tariq Muneer B.Eng. (Hons.), M.S., Ph.D., D.Sc., C.Eng., FCIBSE, Millennium Fellow, is a senior academic who got his engineering education in India, the United States, and Britain. Presently, a Professor of Energy Engineering at Edinburgh Napier University, Scotland, he is a Chartered Engineer who has researched and taught engineering in the United States, North Africa, and Britain for the past 45 years. He has had the privilege of working in four continents and very many countries. He has authored 10 books that were published by Elsevier, Springer, Taylor and Francis, Wiley, and MDPI Press of Switzerland. He has written over 280 articles and supervised 40 doctoral students. Recipient of numerous awards and research grants, he enjoys computer programming to solve mathematical puzzles.

Assoc. Prof. Stoyanka Ivanova Architect, M.S., Ph.D., is currently the Head of the Department of Computer-Aided Engineering at the University of Architecture, Civil Engineering, and Geodesy in Sofia, Bulgaria. She conducts research on computer modeling and its application in the fields of solar architecture, solar energy, and energy efficiency. She has published two books related to solar energy and programming with VBA. She has written over 70 articles and supervised several doctoral students. For decades, she has taught thousands of students to code in numerous programming languages—Fortran, Pascal, C/C++, VBA, VB.NET, etc.

List of MS Excel/VBA Workbooks

Note: unless otherwise mentioned all files have xlsm extension.

Ex01-01   LOOKUP tables (xls file)
Ex01-02   Advanced (Two-dimensional) lookup tables
Ex01-03   Developing Macros by writing your own VBA code
Ex01-04   Generalized code for the class of students
Ex01-05   Obtaining day of the week for a given date
Ex01-06   String operations for fun: Leading on to the game of Scrabble
Ex01-07   Matrix operations (xls file)

Ex02-01   The Scrabble word game
Ex02-02   Word-wheel
Ex02-03   Multiply large numbers
Ex02-04   Computing the exact sum of a long geometric series
Ex02-05   Factorial of a large number
Ex02-06   Wijuko number puzzle: matrix applications (xls file)

Ex03-01   Chauvenet's criterion (xls file)
Ex03-02   Solution of a single nonlinear algebra equation
Ex03-03   Solution of a system of algebra equations: Hooke and Jeeves method
Ex03-04a   Solution for fluid friction factor in a single step (xls file)
Ex03-04b   Solution for fluid friction factor via iterative procedure (xls file)

Ex04-01   RGB definition and use of gray shades
Ex04-02   RGB definition and use of false- and pseudo-colors
Ex04-03   RGB definition and of false- and pseudo-colors with gamma correction
Ex04-04   Pseudo-raster graphics, using colored Excel cells
Ex04-05   Visualization of NASA maps in Excel cells
Ex04-06   Visualization of parts of NASA world maps in Excel cells
Ex04-07   Pseudo-raster graphics using Shape collection
Ex04-08   Combination of pseudo-raster graphics using cells and Shape
            collection
Ex04-09   Drawing with vector primitives
Ex04-10   Combining bitmap graphics with vector graphic
Ex04-11   Distance between two points in a plane or 3D space

Introduction

Puzzles offer many benefits to humans as they develop. From simple puzzles that are based on simple shapes to complex, real-world problems a journey of mental development may be undertaken in an evolutionary process. A relevant discussion on this topic has been provided by Manno, 2013 [https://teach.com/blog/the-benefits-of-puzzles-in-early-childhood-development/]

Mathematical puzzles are then, the next logical process that provide mental challenges and take the subject through a similar development process. Mathematical puzzles offer similar benefits as mentioned above with the added benefits of multiple solutions and potential to analyze the merits and demerits of each solution.

Modern engineering design and analysis demand for increased productivity, reduced costs, faster time to work completion, improved quality, and worldwide access to information. In the past three decades, Information Technology (IT) advances have been a catalyst to huge changes in the way design practices are run. Undoubtedly the key drivers to this change have been the explosive breakthrough in the design and development of integrated circuit chips and the emergence of Computer Aided Design (CAD) software.

The software environment Office 365 is used by over a million companies worldwide. Office 365 is the Microsoft software group which contains among other applications the Excel workbook package. In a typical workbook, there could be any number of worksheets.

Worldwide, Excel is used by an estimated 750 million people with a large number of users in the academic, business, and research sectors.

Computer 'Spreadsheets' have been around for over 50 years. After a brief development stage within the mainframe computer environment, they were quickly adopted by Apple and then DOS-based PCs. When first introduced within the microcomputer world the spreadsheet 'VisiCalc' was an immediate success which was followed by 'Lotus 1-2-3' and then Microsoft (MS) Excel. The introduction of

Supplementary Information The online version contains supplementary material available at (https://doi.org/10.1007/978-3-030-94085-0_1).

T. Muneer and S. Ivanova, *Excel-VBA*,
https://doi.org/10.1007/978-3-030-94085-0_1

Visual Basic for Applications (VBA) computing environment within MS-Excel has provided many benefits for the users. The VBA codes are also referred as '**Macros**'.

There are, of course, over 460 functions that are readily available within Excel which is the first port-of-call by most users.

In this book, a large number of Excel/VBA workbooks have been developed to introduce the reader to start coding on their own. Note that coding has been identified as an essential need at the school level and a great many countries start teaching their school-age children very early on. Of course, occasionally we also introduce to the reader a few key '**Functions**' that are available within Excel. Note that where deficiency is noted in the library of available **Functions**, the reader has the ability to develop their own **Functions**. The latter are referred as **user-defined Functions**.

We shall now proceed to introduce **Functions**, **user-defined Functions** and **Macros**. This task is accomplished by presenting examples that respectively address the above-mentioned techniques.

Example 1.1 LOOKUP tables
Problem definition
In this example, the use of MS-Excel in-built functions is presented.

Tabulated data, in lookup form, is often used by design engineers. Excel provides a useful facility for creating Lookup Tables. Excel's Lookup function is based on the linear interpolation method as described above. Through the lookup facility the user may find one piece of information that is based on another piece of information. A lookup table consists of a column or row of ascending values, called 'compare values', and corresponding data for each compare value. This is demonstrated in sheet 'Properties' of the workbook Ex01-01.xls and also via Table 1.1, which give the thermo-physical properties for air. In this illustration, the first column (temperature of the gas) contains data for the compare values. The corresponding data is the thermo-physical property values.

Most fluid flow and heat transfer problems require an estimation of the thermo-physical properties of a fluid at a certain temperature. Inevitably, this requires interpolations for each of the desired thermo-physical properties.

Consider the flow of air at a temperature of 40 °C. Find all thermo-physical properties provided in the workbook Ex01-01.xls.

Table 1.1 Thermo-physical properties of air expressed as a function of temperature

T (K)	r (kg/m^3)	c_p(kJ/kg K)	μ (N s/m^2)	ν (m^2/s)	K (W/m K)	A (m^2/s)	Pr
100	3.556	1.032	7.11E−07	2.00E−06	0.0093	2.54E−07	0.786
150	2.336	1.012	1.03E−06	4.43E−06	0.0138	5.84E−07	0.758
200	1.746	1.007	1.33E−06	7.59E−06	0.0181	1.03E−06	0.737
250	1.395	1.006	1.60E−06	1.14E−05	0.0223	1.59E−06	0.720
300	1.161	1.007	1.85E−06	1.59E−05	0.0263	2.25E−06	0.707
350	0.995	1.009	2.08E−06	2.09E−05	0.0300	2.99E−06	0.700

Solution

- Open the workbook Ex01-01.xls. This workbook consists of two sheets, namely 'Compute' and 'Properties'.
- Activate the sheet 'Compute' by clicking on its tab. This sheet includes the computations performed by Excel.
- Read the example text carefully and insert the given data for air temperature, 40 °C in cell **B4**.
  - The table of thermo-physical properties for air is provided in cell range **(A11:H45)** of 'Properties' sheet. This sheet is accessed by clicking on its name tab. The air temperatures are given in the far-left column **(A11: A45)** in Kelvin. Therefore, T_a should be $(40 + 273.15 = 323.15$ K$)$ **(cell B6)**. This temperature $(323.15$ **K**$)$ does not exist in the temperature column **(A11:A45)**. In a manual routine, the temperatures immediately below and above T_a, termed T_{down} and T_{up}, are firstly located in the column **(A11:A45)**. Secondly, the properties at T_{down} and T_{up} are determined. Finally, the required properties at T_a are found by interpolating between the corresponding properties at T_{down} and T_{up}. This cumbersome process of calculations can be automated by using the Excel LOOKUP facility. Thus, the required properties are determined automatically each time the value of T_a is changed. The LOOKUP facility is explained below.
    T_{down} **(cell B7)** is found by using the following form of VLOOKUP function:
    =VLOOKUP (lookup-value, table-array, column-index),
    where,
- **lookup-value** is the value to be found in the first column of the table, i.e. the value of **(cell B6)** = 323.15 K.
- **table-array** is the table to be searched, i.e. **(A11:H45)**. (Values of column **(A11:A45)** must be in ascending order).
- **Column-index** is the column number in the table from which the matching value must be returned, e.g. **1** returns the value in the first column (the temperature column).

VLOOKUP searches for a value in the leftmost column of a table and returns a value in the same row from a column you specify in the table. If VLOOKUP cannot find the **lookup-value**, it returns the largest value in the table that is less than or equal to the **lookup-value**. If the **lookup-value** is smaller than the smallest value in the first column, VLOOKUP returns the #N/A error value. In this example, VLOOKUP returns the value of 300 K for T_{down} **(cell B7)**.

The air properties at T_{down} are found in the same way. The temperature 300 K **(cell B7)** is the **lookup-value**. The **column-index** should be changed according to the number of the column of the required property, e.g. 2 for ρ, 3 for c_p, 4 for μ, and so on. For example, to find the value of ρ at 300 K, **(cell C7)** = 1.161 kg/m³. VLOOKUP (B7,A11:A45,2).

To find T_{up}, the table array **(A11:H45)** must be in a descending order as shown in cell range **(A47:H81)**.

Three Excel functions are then used to find the value of T_{up}. These functions are shown in cells **(J11:J13)**.

(cell J11) = MATCH (B6, A47:A81, -1)

This function gets the relative position of the smallest value in the far-left column of the array table **(A47:H81)** that is greater than or equal to the **lookup-value** T_f in **(cell B6)**. In this case, the MATCH function gives a value of 30. This means that T_{up} is located in row number 30 in the array table **(A41:H75)**.

(cell J12) = ADDRESS (J11+46,1)

This function converts the relative position of a cell to an absolute position. In this case, it gives the absolute position for T_{up} in the sheet Properties as (cell A76).

(cell J13) = INDIRECT(J12)

This function gives the value stored in a cell whose address is known. In this case, it gives the value stored in **(cell A76)** which is the required T_{up}.

The properties of air at T_{up} are found by using the VLOOKUP function as explained above in the case of the properties for T_{down}. For example, ρ at 350 K, **(cell C5) = 0.995 kg/m³** VLOOKUP (B5,A11:A45,2).

Finally, the properties of air at T_a are calculated by performing a linear interpolation between the corresponding values at T_{down} and T_{up} , e.g. the value of ρ at 323.15 K,

(cell C6) = IF(C5 = C7, C5, (B6−B7) * (C5−C7)/(B5−B7) + C7)

This expression takes into account the possibility that the **lookup-value** T_a may exist in the lookup-table, i.e. (T_{down} = T_{up}).

Therefore, the properties at T_a are shown in cell range **(C6:I6)**. As a demonstration only the following properties are shown in this example: $v = 1.82 \times 10^{-5}$ m²/s **(cell F6)** and Pr = 0.704 **(cell I6)**.

Discussion

This example used the Functions available within Excel. No programming (coding) was needed. However, what was demonstrated is that a lot of tedium is needed with a large number of Functions used. Overall the number of calculation steps was prohibitive to first obtain thermo-physical properties from the table of ascending and then descending values of temperature. In the next example, we demonstrate the ease with which interpolation can be carried out by writing a simple code in VBA.

The related file, Ex01-01, was saved as a XLS workbook. Workbooks that have connected VBA codes have to be saved with 'xlsm' extensions and that shall be demonstrated in the next example.

Example 1.2 Advanced (Two-dimensional) lookup tables
Problem definition
In this example, a procedure is presented for creating a '**user-defined Function**'.

In Example 1.1, the lookup function was introduced. That particular facility enables linear interpolation of data and searches in one-dimensional tables, i.e. only one independent variable can be handled. In engineering practice, however, particularly in thermo-fluid design, two-dimensional tables are commonly used. For example, in thermodynamics and heat transfer-related problems calculations that are related to steam boiler design and analysis, one encounters thermodynamic and transport properties that are functions of two independent variables, namely, temperature and pressure. Three such property matrices are given in the file Ex01-02.xlsm that are part of this book and available to download from the companion website (https://link.springer.com/chapter/ 10.1007/978-3-030-94085-0_1). That information is shown here via Table 1.2. The matrices include tables for density, enthalpy, and entropy for a steam/water mixture for a common range of temperature and pressure.

Presently, the Visual Basic for Applications (VBA) facility within Excel has been used to develop a two-dimensional search and interpolation routine. Its use is demonstrated via this, Example 1.2.

Solution
Use the VBA program provided within Ex01-02.xlsm to explore interpolation for enthalpy of steam/water mixture. Compute enthalpy for the range of temperature and pressure values given within the 'Calculations' sheet of the above-mentioned Excel file.

Note that the above file has extension as 'xlsm', the 'm' denotes a macro with an in-built user-defined Function. Unless saved with this extension the Function or VBA code will not be saved. That point is worth noting.

By simultaneously pressing the 'Alt' + 'F11' keys display the VBA routine. The function 'Interpolateh(T, P)' developed by the present authors is displayed. This function will linearly interpolate enthalpy values for any given temperature and pressure. The formula given in cell **C5** may be copied down column C to obtain enthalpy values for other temperature and pressure combinations.

Table 1.2 Enthalpy (h) data for water/steam substance

P (MPa)	0.100	0.200	0.500	1.000
T	h	h	h	h
C	kJ/kg	kJ/kg	kJ/kg	kJ/kg
100	2675.1	418.4	418.7	419.0
110	2695.7	460.7	460.9	461.2
120	2715.9	503.0	503.2	503.6

It is only possible to interpolate within one given phase of the substance, i.e. the user may interpolate for compressed liquid or superheated phases.

Discussion
In engineering applications, the user is often required to use property data. The user is often required to perform interpolations. In Examples 1.1 and 1.2, linear interpolation technique was demonstrated via two different procedures, i.e. Example 1.1 used the in-built functions that are available within the MS-Excel environment whereas Example 1.2 dealt with the same task of interpolation by developing a '**user-defined Function**'. Note that Example 1.1 needed only one-dimensional interpolation—the thermo-physical properties were dependent only on temperature. Even then the interpolation procedure became quite involved. On the other hand, Example 1.2 demonstrated a two-dimensional search and interpolation technique using an expressly developed **user-defined Function**. The latter technique amply proved that codes written in VBA language are effective and offer simplicity. In the rest of the book in majority of the cases, **user-defined Functions** and Macros shall be developed and the reader will be encouraged to write their own code, via Exercises at the end of each chapter.

Example 1.3 Developing Macros by writing your own VBA code
In this example, the development of Macros is presented using VBA programing language. We shall attempt this by taking a very simple example.

Problem definition
An engineering teaching program has enrolled students whose marks are provided in Table 1.3. Develop your own VBA code to obtain the average mark for the given class of students.

Solution
The ten-line VBA code is provided in Table 1.3b which has been broken into three segments, the middle segment performing the data input and processing tasks and the last segment generating the desired output. The first and last line of the code is standard lines of code which name the Macro and close it. Note

Table 1.3 Marks awarded to a class of engineering students

Student	Mark awarded
A	57
B	78
C	35
D	41
E	65

Table 1.4 Average mark calculation for the class of engineering students

```
Sub classaverage()
Dim studentmark(5) As Integer
  summark = 0

For i = 1 To 5
      studentmark(i) = Sheets("Main").Cells(i + 1, 2).Value
      summark = summark + studentmark(i)
  Next i

  Sheets("Main").Cells(8, 1).Value = "Class Average"
  Sheets("Main").Cells(8, 2).Value = summark / 5
End Sub
```

that 'Sub' is short for Sub-routine. The second line of the code creates a memory space for storage of student marks and the third line initiates the totaling process.

Note that the code under discussion is presented in MS-Excel workbook Ex01-03.xlsm. Launch that workbook by double-clicking the file name in your PC. Simultaneously click **Alt + F11** keys to open 'Microsoft Visual Basic for Applications' code development and viewing facility. If the code is not already on display then double-click on **Modules,** then **Module1.** You will be able to view the contents of Table 1.4.

There are at least two ways to run the code: (a) simultaneously click **Alt + F8** keys to launch the Macro dialog box. Select the **Macro** (there is only one macro in this Example file), then click **Run,** (b) from the 'Microsoft Visual Basic for Applications' code development and viewing facility, go to Run menu and then **Run Sub/UserForm,** or simply hit the **F5** key.

Discussion
Table 1.4 provided the simplest possible code for the task at hand. There could be any number of variations that would make the code more elegant and versatile but at the cost of added complexity. One such variation that would further generalize the procedure is presented in Example 1.4.

Example 1.4 Generalized mark calculator for a given class
Problem definition
Table 1.4 provided a simple VBA code to calculate the average mark for the class of students. In that code the number of students was fixed to 5. Further

Table 1.5 Generalized code for the class of students

```
Sub classaverage()
Dim studentmark(500) As Integer
Dim studentname As String
     summark = 0
     studnumb = 0

For i = 1 To 500
     studentname = Sheets("Main").Cells(i + 1, 1).Value
     studentmark(i) = Sheets("Main").Cells(i + 1, 2).Value
     If (studentname = "") Then
              GoTo 1
     End If
     summark = summark + studentmark(i)
     studnumb = studnumb + 1
Next i
1:

     Sheets("Main").Cells(8, 1).Value = "Class Average"
     Sheets("Main").Cells(8, 2).Value = summark / studnumb
End Sub
```

develop the VBA code to obtain the average mark for a class that may have as many as 500 students.

Solution

A more generalized code is presented in Table 1.5 and is also included in the digital file Ex01-04.xlsm that is downloadable from this book's website. Now the code contains 18 lines which enables it to handle marks for 500 students and there is no need for the number of students to be specified as the code will determine it by encountering a null character. That is determined in the 9th line of the code. As soon as null character is encountered the code will pass the execution to 10th and then to 15th lines of code.

Discussion

Run this code from Ex01-04.xlsm and explore further by providing more student names and marks. Note that this code can now handle marks for a class of 500 students. Experimentation is a good thing. Try to develop the code furthermore by trying to find the highest and lowest scoring students and the marks they obtained.

Example 1.5 Obtaining day of the week for a given date
Consider the following exercise. You are asked to find out whether the following are leap years?

(a) 2024
(b) 2100.

You may check your calculations against calendars that are available from websites provided on the web.

Problem definition
The common perception is that if the year number is divisible by 4, then it ought to be a leap year, i.e. the year 2024 referred above will be a leap year. However, even though the year 2100 obeys the above rule for leap year determination, it will NOT be a leap year. That is due to the fact that there is another over-riding rule that century years will be leap years provided they are divisible by 400! Since 2100 is not divisible by 400 it will not be a leap year, nor will the years 2200 or 2300 will be. The next century year which will be a leap year will therefore be 2400.

Write a VBA code that, given any date, will enable you to determine the day of the year (known as Julian day) and the day of the week. The Julian days run from 1 to 365, 1 being assigned to January 1st and 365 to December 31st for a non-leap year.

Note that the algorithm for the above task is easily available from many websites, but also from good programming books such as Press et al. (1992) [W Press, S Teukolsky, W Vetterling and B Flannery, Numerical recipes in FORTRAN: The art of scientific computing, Cambridge University Press, Cambridge, UK].

Solution
Open file EX01-05.xlsm which contains three functions which, for a given date, return the following outputs:

- Julian day number.
- Number for the day of the week, Sunday being assigned the value of 0 and Saturday being 6.
- Conversion of the day of the week to the weekday name, i.e. from Monday through to Sunday.

You may open the VBA code by simultaneously clicking Alt + F11 keys. Note that in this example rather than using an executable Macro use of the FUNCTION facility is made. FUNCTIONS have the advantage that they are self-executed. Therefore, as you change the input data the output is obtained without any further interaction.

Study the code carefully and learn how the FUNCTION development facility could be effectively deployed.

Next try this experiment which should address the issue raised at the beginning of this Example. Find the 'Day of the week' for the following dates, the outputs obtained from Ex01-05.xlsm is also shown next to the given dates:

Input	Output
28 February 2024	Wednesday
29 February 2024	Thursday
01 March 2024	Friday

Now try the above dates, but for the year 2100. You will obtain the following output from the file Ex01-05.xlsm.

Input	Output
28 February 2100	Sunday
29 February 2100	Monday
01 March 2100	Monday

Note that 29 February 2100 was ignored as the year 2100 will not be a leap year.

Discussion

In this example, we introduced another VBA tool, i.e. FUNCTION. Macros and FUNCTIONS have their own merits and demerits and their properties will emerge as we progress through the book. This example also provided a useful tool for obtaining the day of the week which is a required information to plan events which would need prior knowledge of the date when clocks are changed with respect to daylight saving time.

Example 1.6 String operations for fun: Leading on to the game of Scrabble

Problem definition

In Chap. 2, we shall demonstrate how the VBA code may be fun and used as a board-game companion for games such as Scrabble. In the present example, we wish to extract all three-lettered words from a given digital English language dictionary. Furthermore, we wish to split the word string into individual letters so that we may compare them against the letters that we hold.

Table 1.6 Code for selecting three-lettered words and splitting them into individual letters

```
Sub Allwords()
Dim strlnt As Integer
Dim bword, cword, frstchr, bmfc(3) As String
    strlnt = 3

    j = 1
    For i = 1 To 58112
        bword = Sheets("Dictionary").Cells(i, 1).Value
        cword = bword
        wrdlnt = Len(bword)

        If (wrdlnt = strlnt) Then
            Sheets("Main").Cells(j, 24).Value = bword
            ncode = 0
            For M = 1 To strlnt
                frstchr = Left(cword, 1)
                bmfc(M) = frstchr
                cword = Right(cword, (strlnt - M))
                Sheets("Main").Cells(j, 24 + M).Value = frstchr
            Next M
            j = j + 1
        End If
    Next i
End Sub
```

Let us say a player is holding three letters in his/her hand, the letters being 'A', 'R', and 'W'. We wish to find out how many words have those three letters. From hindsight we know that there are only two such words—'raw' and 'war'.

Solution
Launch the Excel file 'Ex01-06.xlsm' and open the code by simultaneously clicking Alt + F11 keys. The code displayed in Table 1.6 has only 20 lines of code. Study this code carefully, the salient points being:

- The 58,112 words within the companion dictionary are read in a loop, the words being viewable in the like-named worksheet of the file Ex01-06. xlsm.

- The command 'wrdlnt = Len(bword)' finds the length of the string and if that length is three-letter long then the VBA code proceeds further to split the word into it constituent letters.
- The split letters are then written as output in worksheet called 'Main'.

Discussion
A very simple code was developed for string operations. This code shall be further developed to develop applications in board games such as Scrabble, a more robust introduction being provided in Chap. 2.

There are very many applications of string applications within the word processing and texting that is available in mobile smart phones. Three such applications that directly result from such an exercise and which can easily be identified are:

- Word prediction: as the user starts to key-in a few letters of the desired word the in-built software predicts the likely outcomes. The more letters you key-in the narrower the search becomes.
- If the letters are jumbled-up in the wrong manner, the software can decipher the correct word. That is easily manageable as shall be shown in Chap. 2.
- The word prediction can easily be taken to the next stage, i.e. sentence prediction. That can be managed by accessing paragraphs from standard texts.

Example 1.7 Matrix operations: Leading on to the number games
Problem definition
Consider the following simple system of simultaneous algebra equations of two unknowns: X and Y,

$$X + Y = 3$$
$$2X + Y = 4$$

The solution is X = 1 and Y = 2.
Obtain the solution to this problem via matrix algebra.

Solution
Let us attempt to solve this using matrix operations. We may write the above system in matrix form as shown in Eq. (1.1):

$$\begin{bmatrix} 1 & 1 \\ 2 & 1 \end{bmatrix} \begin{bmatrix} X \\ Y \end{bmatrix} = \begin{bmatrix} 3 \\ 4 \end{bmatrix} \tag{1.1}$$

Note that the square matrix on the left is the 'coefficient' matrix which is followed by the matrix containing the 'unknowns' column matrix. The column matrix on the right-hand side contains the 'constants'.

The solution to this problem may be obtained thus—Eq. (1.2),

$$\begin{bmatrix} X \\ Y \end{bmatrix} = \text{INVERSE} \begin{bmatrix} 1 & 1 \\ 2 & 1 \end{bmatrix} \cdot \begin{bmatrix} 3 \\ 4 \end{bmatrix} \tag{1.2}$$

The above matrix algebra is easily handled in MS-Excel and shall be demonstrated in the present example.

Launch the file Ex01-07.xls which contains the three matrices of Eq. (1.1). The given 'coefficient' and 'constants' matrices are shown in yellow-colored cells while the yellow-colored cells contain the 'unknowns' and the 'solution' column vectors.

There are only two computational steps:

1. Select cells C2:D3 using your computer mouse, then enter the formula MINVERSE(A2:B3). Then while simultaneously holding the CTRL and Shift keys hit the Enter key. You will thus obtain the inverse matrix.
2. Select cells H2:H3 using your computer mouse, then enter the formula MMULT(C2:D3, F2:F3). Then simultaneously holding the CTRL and Shift keys hit the Enter key. You will thus obtain the product of the two matrices shown in the right-hand side of Eq. 1.2. The solution may be seen in cells H2:H3.

Discussion

Matrix algebra is a powerful tool in solving simultaneous equations which are encountered in numerical puzzles as well as engineering design and analysis. We shall explore this further in the next chapter.

1.1   Conclusion

In this chapter, we have introduced simple applications of Excel/VBA such as one- and two-dimensional tables, code building for tasks such as obtaining averages for a class of students, string operations with the view to apply them for development of computer games, and matrix operations for solving numerical puzzles such as those found in newspapers. The limitations of functions available in MS-Excel were demonstrated in contrast to the logics prowess that is available via VBA.

Exercises

E1.1   Which word has three consecutive pairs of letters? Find out via manipulation of the VBA code that was provided in this chapter.

Answer: Bookkeeper is the word with 3 pairs of consecutive letters.

E1.2   Write an MS-Excel/VBA macro to determine the last Saturday in October
for any given year. Find out the last Saturday in October for the years 2028
and 2100. Note that the significance of the last Saturday in October is that it
determines the change of clocks back to GMT in the United Kingdom.

Hint: You will need to alter the code provided for obtaining the day of the
week so that it runs iteratively, while 'looking' for Saturday'.

E1.3   Refer to Ex01-06.xlsm and worksheet 'Main'. The three given letters are
provided in cells B1:D1. The code has successfully split the letters of all
three-lettered words and these are recorded in columns Y, Z, and AA.
Extend this code further so that only those three letters that match the given
letters of cells B1:D1 are recorded in columns Y, Z, and AA.

Note that this problem shall be explored in further detail in Chap. 2.

Games and Puzzles

2

In this chapter, we shall explore the fun way in which Mr games to the next level and learn algorithm building in the process.

Example 2.1 The Scrabble word game

Problem definition

Scrabble is a word game in which two to four players score points by placing tiles on the game board, each tile bearing a single letter. The game board is divided into a 15 × 15 grid of squares. The tiles must form words that, in crossword fashion, read left to right in rows or downward in columns, and be included in a standard dictionary. One good dictionary is the Oxford English Dictionary, also known as OED.

The game is available for sale in very many countries and there are several hundred million players and thousands of Scrabble clubs around the world. The game progresses as players' form words on the board with the total of word-forming score being the numerical sum of the constituting tiles, each tile having a unique score. For example, with reference to Fig. 2.1 if the word formed is SPIRE, then the word-score will be $1 + 3 + 1 + 1 + 1 = 7$.

Using the tiles shown in Fig. 2.1 write a VBA code that may form as many words as possible.

Solution

Refer to Fig. 2.1 which shows the tiles picked by the four Scrabble players: **a** (top-left), **b** (top-right), **c** (bottom-left), and **d** (bottom-right).

We need to write a VBA code that achieves the following tasks. It is assumed that readers will have a preliminary knowledge of VBA programming language as children in schools are now taught coding even in earlier years of their education.

Supplementary Information The online version contains supplementary material available at (https://doi.org/10.1007/978-3-030-94085-0_2).

Fig. 2.1 The Scrabble board and alphabet tiles picked by one player

- Read the letters for any given player
- Sequentially read all the Z-lettered words in a given dictionary, where Z takes the value of 2, 3, 4, 5, 6 and 7.
- Sequentially match the words from letters provided

The code is displayed in Table 2.1 with line numbers identifying each line of the code and blocks of program separated out. Explanation of the code is provided here.

Lines 1–10 contain the main part of the program with the word-forming sub-routine being called in line# 6. Note that the loop from lines 4–7 instructs the machine for sequentially forming words with 2, 3, 4, 5, 6, and 7 words respectively. The word-forming sub-routine starts at line # 11.

Discussion

This exercise was handed over to a number of players of varying age and professional background—a first-year, high school student to a retired business executive and even an English language teacher. Table 2.2 shows the score achieved by those volunteers. Those scores are compared to the present software's performance.

At this stage, the reader is encouraged to execute the code shown in Table 2.1 and get a list of the words formed by the software.

Note that in the Exercises section we shall revisit the application of Example 2.1 for more rigorous applications of word formation. In that respect we refer to Figs. 2.2 and 2.3 wherein letter tiles for four players, and a blank tile that may be used as a wild card are respectively shown.

Table 2.1 VBA code for Example 2.1: Scrabble word game

```
1     Sub Allwords()
2     Dim strlnt, iwrdlnt As Integer
3     jfinal = 1
4     For iwrdlnt = 2 To 9
5     strlnt = iwrdlnt
6     GoSub 10001
7     Next iwrdlnt
8     Sheets("Main").Select
9     Range("A1").Select
10    End

11    10001 'Subroutine for word creation
12    Dim amfc(22), bmfc(22), singlet(26) As String
13    Dim aword, bword, cword, dword, eword(10000), frstchr As String
14    Dim sglet, sdlet, declet(26), vallet(26) As Integer
15    Dim givenlet(26), foundlet(26) As Integer
16    Dim fglet(26), fdlet(26)
17    NgivenLet = Sheets("Main").Cells(4, 2).Value
18        Sheets("dump").Select
19        Range("A2:BA2").Select
20        Selection.ClearContents
21        Range("A1").Select
22        Sheets("dump").Select
23        Range("A5:BA500").Select
24        Selection.ClearContents
25        Range("A1").Select

26    For km = 1 To 26
27    declet(km) = Sheets("Dictionary").Cells(km, 2).Value
28    vallet(km) = km
29    fglet(km) = 0
30    fdlet(km) = 0
31    Next km

32    For km = 1 To NgivenLet
33    singlet(km) = Sheets("Main").Cells(1, km + 1).Value
34    Next km

35    For kl = 1 To 9
36    sumuplet = 0
37    uplet = singlet(kl)
38    For kn = 1 To 26
39    If (uplet = declet(kn)) Then
40    Sheets("dump").Cells(2, kl).Value = vallet(kn)
41    End If
42    Next kn
43    For km = 1 To 9
44    If (uplet = singlet(km)) Then
45    sumuplet = sumuplet + 1
46    End If
47    Next km
48    Sheets("dump").Cells(2, kl + 26).Value = sumuplet
49    Next kl
```

(continued)

Table 2.1 (continued)

```
50    j = 1

51    For i = 1 To 58112
52    bword = Sheets("Dictionary").Cells(i, 1).Value
53    cword = bword
54    wrdlnt = Len(bword)
55    If (wrdlnt = strlnt) Then
56    ncode = 0

57    For M = 1 To strlnt
58    frstchr = Left(bword, 1)
59    bmfc(M) = frstchr
60    bword = Right(bword, (strlnt - M))
61    Next M

62    For M = 1 To strlnt
63    mcode = 0
64    For km = 1 To 9
65    If (bmfc(M) <> singlet(km)) Then
66    mcode = mcode + 1
67    End If
68    Next km
69    If (mcode = 9) Then
70    GoTo 301
71    End If
72    Next M

73    Sheets("dump").Cells(j + 4, 24).Value = cword
74    eword(j) = cword
75    For kl = 1 To strlnt
76    sumuplet = 0
77    uplet = bmfc(kl)

78    For kn = 1 To 26
79    If (uplet = declet(kn)) Then
80    Sheets("dump").Cells(j + 4, kl).Value = vallet(kn)
81    End If
82    Next kn
83    For km = 1 To strlnt
84    If (uplet = bmfc(km)) Then
85    sumuplet = sumuplet + 1
86    End If
87    Next km

88    Sheets("dump").Cells(j + 4, kl + 26).Value = sumuplet
89    Next kl

90    j = j + 1
91    End If
92    301
93    Next i
```

(continued)

Table 2.1 (continued)

```
94    sumwords = j - 1
95    For kl = 1 To 26
96    givenlet(kl) = 0
97    foundlet(kl) = 99
98    Next kl

99    For km = 1 To 9
100   wallet = Sheets("dump").Cells(2, km + 1).Value
101   givenlet(wallet) = Sheets("dump").Cells(2, km + 26).Value
102   Next km

103   For kn = 1 To sumwords
104   For km = 1 To strlnt
105   icode = 0
106   wallet = Sheets("dump").Cells(kn + 4, km + 1).Value
107   foundlet(wallet) = Sheets("dump").Cells(kn + 4, km + 26).Value
108   Next km
109   Next kn

110   For nk = 1 To NgivenLet
111   sglet = Sheets("dump").Cells(2, nk).Value
112   fglet(sglet) = Sheets("dump").Cells(2, 26 + nk).Value
113   Next nk

114   For kn = 1 To sumwords
115   For km = 1 To strlnt
116   sdlet = Sheets("dump").Cells(kn + 4, km).Value
117   fdlet(sdlet) = Sheets("dump").Cells(kn + 4, km + 26).Value
118   If (fdlet(sdlet) > fglet(sdlet)) Then
119   Sheets("dump").Cells(4 + kn, 23).Value = "X"
120   End If
121   Next km
122   Next kn

123   For kn = 1 To sumwords
124   codeX = Sheets("dump").Cells(kn + 4, 23).Value

125   cword = Sheets("dump").Cells(kn + 4, 24).Value
126   If (codeX = "") Then
127   Sheets("Main").Cells(jfinal, 24).Value = cword
128   jfinal = jfinal + 1
129   End If
130   Next kn
131   Return
132   End Sub
```

Note that the file Ex02-01.xlsm contains the above VBA code

Table 2.2 Word-forming scores achieved by Scrabble players

Player	Number of words formed
1	27
2	35
3	34
4	35
5	19
Software	65

Fig. 2.2 Alphabet tiles picked by four Scrabble players

Fig. 2.3 The Scrabble board and alphabet tiles picked by one player. Note that the second tile from left is blank as it is a wildcard. That tile may be used in place of any letter thus increasing the potential to get a higher word-forming score

Example 2.2 Word-wheel

Problem definition

Newspapers around the world carry games and puzzles on their back pages. One such game is called 'word-wheel'. A typical example of such word-wheel is shown in Fig. 2.4.

The game player is asked to form words that contain three or more letters. The condition is that each word must contain the central letter which is 'A' in this case and letters must be used only once. There will be one nine-letter word, constituting of all the letters shown in the figure.

Let us start nice and simple. We can easily form words such as 'tan' and 'taunt'. Using your memory, try to make as many words as possible and keep a paper record of that. Now let us see how we can use Excel–VBA.

Solution

Open the file Ex02-02.xlsm and launch the VBA code by simultaneously clicking Alt + F11 keys. Study the code carefully and compare it against the code of Example 2.1. (Ex02-01.xlsm). There is a great deal of similarity between the two sets of code, the only difference being in the present code use has to be made of the central character which is 'A'. In the VBA code of Ex02-02.xlsm search for the string 'esschar' which contains the central character in its memory. You will notice that the latter string is used in only two places in the present code. We have thus managed to use the code of previous example and convert it for the purpose of solving the 'word-wheel' example.

Discussion

Refer to the Discussion of Example 2.1. Once again the present exercise was handed over to the referred players. Table 2.3 shows the scores achieved by the volunteers. Those scores are, once again, compared to the present software's performance.

The reader is encouraged, once again, to execute the code shown in Table 2.1 and get a list of the words formed by the software for the word-wheel shown in Fig. 2.4.

Table 2.3 Word-forming scores achieved for word-wheel puzzle

Player	Number of words formed
1	19
2	17
3	27
4	42
5	37
Software	57

Fig. 2.4 A word-wheel

Example 2.3 Multiply large numbers

Problem definition

Given two large integer numbers, each of which contains a large number of digits, write a program to compute the exact product of the two numbers.

Solution

The problem may be analyzed by placing ourselves in the place of computer. As an example, let us consider the multiplication of the following two numbers from first principles,

$$
\begin{array}{r}
68 \\
\times 74 \\
\hline
272 \\
476 \\
\hline
5032
\end{array}
$$

The above procedure may be used to multiply numbers of any magnitude by the following steps:

1. Read and save in a dimensional space the two numbers, the multiplicand and the multiplier.
2. Select the first digit of the multiplier.
3. Multiply the above single-digit number with all of the digits of the multiplicand starting with the extreme right-hand digit. Do take care of the carryover and store all the product digits in a two-dimensional array.
4. Keep repeating step 3 until all digits of multiplicand have exhausted.

5. Add the digits in the two-dimensional array taking care of the shift to the left required between two consecutive rows, as shown in above demonstration.

As a trial let us attempt the following problem in which a seven digit number is multiplied with another seven digit number:

Multiplicand:	9 865 438
Multiplier:	7 869 456
Product:	77 635 630 261 728

Launch the file Ex02-03.xlsm and view the VBA code contained within it. Key in the multiplicand and the multiplier as shown in the worksheet named 'Multiply'. Click the 'Multiply Large Numbers' form control button and see the product displayed. Note that any other worksheet not expressly mentioned is only used as a transition tool and hence got to be ignored.

Discussion

The companion VBA code has three main components: Read data from the two strings containing the multiplicand (cell F2) and the multiplier (cell F3), carry out the multiplication process as discussed above, and then return the output in cell F4. Note that the large numbers are entered as string variables which are then converted to numeric. Likewise, the output goes through a reverse procedure. The numerical data for the multiplicand and the multiplier is dumped in the sheet 'Multiplyback' as a vector to overcome the problem of limited width of any given cell. Likewise, the product is also obtained in the latter sheet as vector. The macro can handle numbers of any number of digits. The authors have tried numbers that have over hundred digits. The marco will handle the task without any difficulty. For very large numbers one may also read the numbers from text files.

Study the entire code carefully as it will form the basis of the next two examples.

The above problem was presented by one of the authors Muneer (1991).

Example 2.4 Story of invention of chess

Problem definition

You must have heard the story of invention of chess. An Indian king desired a game which was not to be based on chance. An inventor presented the game of chess to the king. The king was so overjoyed that he offered the inventor half of the kingdom as gift. However, the inventor asked the king for 'only' enough grain to fill the chessboard in such a manner that the first square of the chessboard to have one grain of wheat, the second two, the third four, and so on.

It has been reported that the famous Persian mathematician and astronomer, al-Biruni dealt with the above problem in his time. His answer is exactly the same as that has been presented above (Reference E. Masood, Science and Islam: A history, Icon books, London 2009).

Write a VBA code that will enable you to compute the 'exact' number of grains of wheat that would be required to fill all the 64 squares of the chessboard.

Solution

The number of grains to be allocated to the 64 squares of the chessboard will increase in a geometric progression, thus:

$$1, \ 2, \ 2^2, \ 2^3, \ 2^4, \ 2^5, \ \ldots\ldots, \ 2^{63}$$

The sum of this series may easily be shown to be:

$$\text{Sum} = \frac{2^{64} - 1}{2 - 1} = 2^{64} - 1$$

The exact value of above 'Sum' is 18,446,744,073,709,551,616 −1 or 18,446,744,073,709,551,615. The latter figure is obtained from the code provided in file Ex02-04.xlsm and then '1' taken away from it. That code is also provided in Table 2.4. That number is so large that it is not in general currency, i.e. it is eighteen quintillion, four hundred and forty-six quadrillion, seven hundred and forty-four trillion, seventy-three billion, seven hundred and nine million, five hundred and fifty-one thousand, and six hundred and fifteen.

The above number dwarfs the total of the entire world's economy, which in year 2019, stood at a paltry figure of 87.8 trillion US dollars.

Discussion

The VBA code provided in Table 2.4 is easy to follow and has been split in five blocks. An explanation of those blocks is now provided.

Block 1: Dimension space is created in computer random access memory (RAM) to store each individual integer number for the Sum of the above geometric series. The value of 'exponent' (64) is read from the only worksheet in the workbook from cell E1. Note that the present code is of a general nature and you may also obtain 3 or 4 or 5 or any number to whatever power you choose.

Block 2: The value of IA register is stored as 0. Those zeros will be replaced by the significant integers once 2^{64} is obtained.

Block 3: This is the key block which multiplies 2 by itself in a sequential manner 64 times over and the individual integers of the answer are stored in the IB register.

Block 4: All integer digits from IB register are copied over to IA register.

Block 5: The output is generated in the worksheet, one integer at a time.

Table 2.4 Computing the exact sum of a long geometric series

```
Sub twopower()

Dim IA(65), IB(65)
Dim exponent As Integer
    exponent = Sheets("two_powers").Cells(1, 5) + 1
    IA(1) = 1

    For i = 2 To exponent
        IA(i) = 0
    Next i

    For j = 2 To exponent
        IB(1) = IA(1) * 2 - 10 * Int(IA(1) * 0.2)
        For i = 2 To exponent
            IB(i) = IA(i) * 2 + Int(IA(i - 1) * 0.2)
            IB(i) = IB(i) - 10 * Int(IB(i) * 0.1)
        Next i

        For i = 1 To exponent
            IA(i) = IB(i)
        Next i
    Next j

    For k = 1 To exponent
        Sheets("two_powers").Cells(k, 1) = IA(exponent + 1 - k)
    Next k
End Sub
```

Note that if higher powers of number 2 need to be obtained then two changes are required, i.e. change the value of new exponent cell E1 and change the 'Dim' statement in line 2 of the code to n + 1 where n is the value of the new exponent.

Note furthermore that all leading zeros in column A of worksheet 'two_powers' are to be ignored. The significant digits are provided as output in cell range A46:A65.

Example 2.5 Factorial of a large number

Problem definition

The factorial of a positive integer n, denoted by n!, is the product of all positive integers less than or equal to n.

The factorial operation is encountered in many areas of mathematics, notably in algebra, probability theory, and mathematical analysis. In its basic

Table 2.5 Factorials for numbers from 1 to 200

Number	Factorial
1	1
2	2
3	6
4	24
5	120
6	720
7	5040
8	40320
9	362880
10	3628800
11	39916800
12	479001600
13	6227020800
14	87178291200
15	1307674368000
16	20922789888000
17	355687428096000
18	6402373705728000
19	121645100408832000
20	2432902008176640000
25	1.551E+25
50	3.041E+64
70	1.198E+100
100	9.3326E+157
150	5.713E+262
200	#NUM!

use factorial n counts the possible distinct sequences, i.e. the permutations of n events or possibilities.

It is claimed that the use factorials can be dated to the Talmudic period when they were used to count permutations. Indian mathematicians were also known to make use of them in the twelfth century.

Written by Abraham de Moivre, The Doctrine of Chances: A Method of Calculating the Probabilities of Events in Play was an early work in probability theory. It first appeared in Latin in 1711, with the first English edition published in 1718. The cover of the latter work is shown in Fig. 2.5.

Table 2.5 presents factorials of numbers that go up to 150. These were calculated by the built-in-function that is available in MS-Excel, namely 'Fact

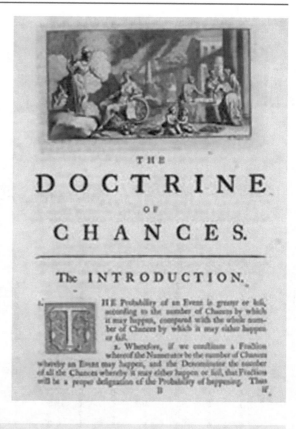

Fig. 2.5 The title page of the 1738 English edition of the book by Abraham de Moivre. *Credit* https://www.maa.org/press/periodicals/convergence/mathematical-treasure-abraham-de-moivres-doctrine-of-chances

(number)'. Note however that once you reach a large number such as 200, Excel fails to deliver and returns the message of '#NUM!'.

We wish to obtain 200! By writing a VBA code which sequentially multiplies $2 \times 3 \times 4 \times 5$ and so on up to 200.

Solution

Open the file Ex02-05.xlsm and launch the VBA code by simultaneously clicking the ALT + F11 keys. The code is very simple and displayed in Table 2.6. Once executed the code will generate a column vector of all the 275 digits that contain the answer for 200!. Note that the initial trail of zeros in the sheet 'Factorial' is to be ignored. The first non-zero digit is written in cell A126 and thereafter the remaining 374 digits follow downwards, right up to cell A500.

The code shown in Table 2.6 is split in three parts. The first and the last parts, respectively, shown in italics carry out the input and output tasks. The middle part of 8 lines of code, shown in bold letters, is the main part of the program that does the number crunching. The code is remarkably simple and concise. The answer thus obtained for factorial of 200 is shown in Fig. 2.6.

Table 2.6 VBA code to obtain factorial 200

```
Sub factorial()
Dim A(500)
A(500) = 1
    For B = 200 To 2 Step -1
        C = 0
        For i = 500 To 1 Step -1
            V = C + B * A(i)
            C = Int(V / 10)
            A(i) = V - 10 * C
        Next i
    Next B

    For i = 1 To 500
        Sheets("Factorial").Cells(i, 1) = A(i)
    Next i
End Sub
```

7886	5	7	8	6	7	3	6	4	7	9	0
5	0	3	5	5	2	3	6	3	2	1	3
9	3	2	1	8	5	0	6	2	2	9	5
1	3	5	9	7	7	6	8	7	1	7	3
2	6	3	2	9	4	7	4	2	5	3	3
2	4	4	3	5	9	4	4	9	9	6	3
4	0	3	3	4	2	9	2	0	3	0	4
2	8	4	0	1	1	9	8	4	6	2	3
9	0	4	1	7	7	2	1	2	1	3	8
9	1	9	6	3	8	8	3	0	2	5	7
6	4	2	7	9	0	2	4	2	6	3	7
1	0	5	0	6	1	9	2	6	6	2	4
9	5	2	8	2	9	9	3	1	1	1	3
4	6	2	8	5	7	2	7	0	7	6	3
3	1	7	2	3	7	3	9	6	9	8	8
9	4	3	9	2	2	4	4	5	6	2	1
4	5	1	6	6	4	2	4	0	2	5	4
0	3	3	2	9	1	8	6	4	1	3	1
2	2	7	4	2	8	2	9	4	8	5	3
2	7	7	5	2	4	2	4	2	4	0	7
5	7	3	9	0	3	2	4	0	3	2	1
2	5	7	4	0	5	5	7	9	5	6	8
6	6	0	2	2	6	0	3	1	9	0	4
1	7	0	3	2	4	0	6	2	3	5	1
7	0	0	8	5	8	7	9	6	1	7	8
9	2	2	2	2	2	7	8	9	6	2	3
7	0	3	8	9	7	3	7	4	7	2	0
0	0	0	0	0	0	0	0	0	0	0	0
0	0	0	0	0	0	0	0	0	0	0	0
0	0	0	0	0	0	0	0	0	0	0	0
0	0	0	0	0	0	0	0	0	0	0	0

Fig. 2.6 Factorial 200 has 375 digits. All of those 375 digits are shown in this figure

Discussion

By writing the output for factorial 200 as a column vector we have solved the problem of displaying large integer numbers. One may always take the column vector and manipulate it in Excel, MS Word, or Notepad applications to display in the manner shown in Fig. 2.6.

Note that if factorial of a larger number needs to be obtained then four changes are required, i.e. in lines 2, 4, 6, 12 of the above code.

Note furthermore that all leading zeros in column A of worksheet 'Factorial' are to be ignored. The significant digits are provided as output in cell range A126:A500.

Example 2.6 Wijuko number puzzle (Matrix applications)
Problem definition
A number of daily newspapers carry numerical puzzles. Among them one that is popular is *Wijuko* or *Sujiko*. Refer to Fig. 2.7 it shows nine squares of which 3 have been filled with numbers. Your task is to fill-in the remaining 6 squares with numbers ranging between 1 and 9 so that the totals for any given set of four squares add up to the circle within the center of those squares.

Solution
These types of puzzles are easily handled via matrix operation. For the sake of analysis, let us number each of the above-referred squares. This is shown in Fig. 2.8.

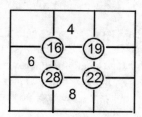

Fig. 2.7 Wijuko/Sujiko puzzle

Fig. 2.8 Preparation to construct solution matrix

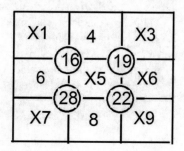

Fig. 2.9 Wijuko/Sujiko puzzle with six missing numbers

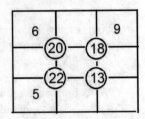

We are now in a position to write the respective set contained in Eqs. 2.1, $X1 + 4 + X5 + 6 = 16$, hence Eq. (2.1)

$$X1 + X5 = 6 \tag{2.1}$$

Likewise, we may write Eq. (2.2)

$$X3 + X5 + X6 = 15 \tag{2.2}$$

$$X5 + X7 = 14 \tag{2.3}$$

$$X5 + X6 = 14 - X9 \tag{2.4}$$

$X1 + X3 + X5 + X6 + X7 =$ SUM OF NUMBERS FROM 1 TO 9 –SUM OF 4,6,8 –X9 which reduces to,

$$X1 + X3 + X5 + X6 + X7 = 27 - X9 \tag{2.5}$$

The reason we have written X9 on the right-hand side is that we have only 5 equations and 6 unknowns. Hence, the solution strategy will be to assume a value for X9, then solve for the remaining unknowns, then check whether the solution makes sense. Otherwise, we will change the assumed value of X9 iteratively. Note that the iterations can be performed manually or via a routine.

We shall first attempt manual iterations.

Launch Excel/VBA file Ex02-06.xls. Note that this file does not have the 'xlsm' extension as it does not have any VBA Macro. Using what we learnt via file Ex01-07.xls carefully compare Eq. 2.1 against the matrices A3:E7 and M3:M7. We are retracing exactly the same steps that we covered in Ex01-07. xls. The 'guess' value for X9 is provided in cell B2 and that will alter cells M6 and M7. Try changing the X9 value of cell B2 from 1 through to 9. You will find that the only value which gives sensible results in the SOLUTION column O3:O7 is 2. Hence X9 will be 2 and the rest of the numbers X1, X3, X5, X6, and X7, respectively, are, 1, 3, 5, 7, and 9.

Discussion
We have extended the lesson of Example 1.7 to good effect to solve a numerical puzzle. Matrix algebra is a powerful tool to handle simultaneous equations and that was presently demonstrated.

2.1   Conclusion

In this chapter, we have furthered the development of advanced applications of Excel/VBA for word-forming and number crunching such as obtaining the exact product of very large numbers, factorial of a large number, or raising an integer number to a large exponent. The examples presented were based on historical narration. More advanced matrix operations were also introduced for solving advanced numerical puzzles.

Exercises

E2.1   A 'heterogram' is a word that is made up of non-repeating alphabets, i.e. every alphabet in its construction is only used once. Examples of five, seven, and ten-lettered heterogram are 'brick', 'stumped', and 'blacksmith'. Using the skills developed in Sections 2.1 and 2.2 list all ten-lettered heterograms that are to be found in the dictionary provided with Ex02-01.xlsm and Ex02-02.xlsm. You need to write a VBA code to undertake this task.

E2.2   Following on from E2.1, collect in a text file all words that end with the letters 'ion'.

E2.3   Refer to Example 2.1 and Fig. 2.1. A VBA code was provided in digital file Ex02-01.xlsm. Develop your own code further so that you obtain the score for each of the words that were formed in the latter example.

E2.4   Refer to Example 2.1 and Fig. 2.2. A VBA code was provided in digital file Ex02-01.xlsm. Develop your own code further so that you obtain the score for each of the words that each of the four players, A, B, C, and D can form. Get the total score for each word thus formed. You may be in a position to find which player has the potential to get the highest score.

E2.5   Refer to Example 2.1 and Fig. 2.3. A VBA code was provided in digital file Ex02-01.xlsm. Develop your own code further so that you obtain the score for each of the words that could be formed using the blank tile. Note that the blank tile is a wildcard. That tile may be used in place of any letter thus increasing the potential to get a higher word-forming score. Table 2.7 provides the points associated with each of the Scrabble letters.

E2.6   A large number of books whose copyright has expired are now available on the web from a number of websites. One such landmark book that is available from Project Gutenberg is 'The Arabian Nights Entertainments

Table 2.7 Points worth of each letter in Scrabble word game

A	1	J	8	S	1
B	3	K	5	T	1
C	3	L	1	U	1
D	2	M	3	V	4
E	1	N	1	W	4
F	4	O	1	X	8
G	2	P	3	Y	4
H	4	Q	10	Z	10
I	1	R	1		

Complete'. That eBook is for the use of anyone anywhere at no cost and with almost no restrictions whatsoever. You may copy it, give it away, or re-use it under the terms of the Project Gutenberg License included with this eBook or online at www.gutenberg.net. The book posted on January 24, 2009 [EBook #5668] has the character set encoding: ISO-8859-1.

Download one such book and then write a VBA code that will enable you to predict words that follow commonly used words. That code may read two consecutive words, one at a time, search for one of your chosen common word and then provide the predicted word that follows your chosen word.

E2.7 Create a text file that contains a large number (say, 20) of 50 digit numbers. Write a VBA code to obtain the sum of those numbers. You may use lessons learnt from Example 2.3.

E2.8 Four Bridge players are enjoying a playing session. Work out the probability of each player getting just one suite of cards, i.e. player one gets all spades, player two, all diamonds, and so on. Obtain the exact answer of the probability using file Ex02-03.xlsm and lesson learnt from Example 2.3.

E2.9 Refer to Example 2.6 in which the Wijuko/Sujiko puzzle was introduced. Obtain the solution for the puzzle whose diagram is shown in Fig. 2.9. The diagram has six missing numbers. Fill-them up with single-digit integer numbers in the squares provided such that the sums of the numbers circumscribing the circles add up to the numbers within those circles. On completion of the puzzle, the nine squares ought to contain numbers 1 through to 9.

Reference

T. Muneer, J. Usher, Teaching computer programming skills: a stimulating approach. J. Scottish Math. Council 45–52 (1991)

Numerical Analysis

3

3.1 Error Analysis Using Chauvenet's Technique

Scientists and engineers are constantly engaged in performing experiments in which data collection is undertaken. The data may either be collected manually or via data-loggers. During the data collection phase errors creep into all experiments regardless of care taken. This action may lead to elimination of erroneous data but that must be consistent and not dependent upon whims or bias on what 'ought to be'. One technique that may be used is the Chauvenet's criterion to reject a datum.

The idea behind Chauvenet's criterion is to find a probability band, centered on the mean of a normal distribution that should reasonably contain all n samples of a dataset. By doing this, any data points from the 'n' data samples that lie outside this probability band can be considered to be outliers, removed from the dataset, and a new mean and standard deviation based on the remaining values and new sample size can be calculated. This identification of the outliers will be achieved by finding the number of standard deviations that correspond to the bounds of the probability band around the mean (d_{max}) and comparing that value to the absolute value of the difference between the suspected outliers and the mean divided by the sample standard deviation. The steps needed are:

- Obtain arithmetic mean, Xm & Standard deviation, SD of the data 'X' where the sample size of X is 'n' data points.
- Obtain deviation for each data point, di = Xi – Xm.
- Using Chauvenet's criterion, test the data for any inconsistencies. The dictum of Chauvenet's test is that a data point max be considered to be suspect if the following condition is met:

Supplementary Information The online version contains supplementary material available at (https://doi.org/10.1007/978-3-030-94085-0_3).

T. Muneer and S. Ivanova, *Excel-VBA*,
https://doi.org/10.1007/978-3-030-94085-0_3

$$n * \mathrm{erfc}(\mathrm{abs}(di)/\mathrm{SD}) < 0.5$$

These steps are demonstrated via Example 3.1.

Example 3.1 Quality control of data using Chauvenet's criterion

Problem definition

Ten readings were taken of the length of an object and these are presented in the second column from left-hand side (LHS) of Table 3.1. Using Chauvenet's criterion perform quality control of the data.

Solution

The first step is to prepare the third and fourth columns of Table 3.2 using mean of the data, $Xm = 5.613$ and the standard deviation, 'SD' of the list of di.

Table 3.1 Length of a given object as recorded by 10 volunteers

Reading no	Length of the object, X
1	5.3
2	5.73
3	6.77
4	5.26
5	4.33
6	5.45
7	6.09
8	5.64
9	5.81
10	5.75

Table 3.2 Analysis of the data presented in Table 3.1 regarding any suspect readings

Reading no	Length of the object, X	di	abs(di)/ SD	n*erfc(abs(di)/ SD)	Chauvenet's test
1	5.3	−0.313	0.527	4.565	
2	5.73	0.117	0.197	7.807	
3	6.77	1.157	*1.946*	0.059	Reject
4	5.26	−0.353	0.594	4.010	
5	4.33	−1.283	*2.158*	0.023	Reject
6	5.45	−0.163	0.274	6.982	
7	6.09	0.477	0.802	2.564	
8	5.64	0.027	0.045	9.488	
9	5.81	0.197	0.331	6.393	
10	5.75	0.137	0.230	7.445	
	Xm	SD			
	5.613	0.5944			

The value of SD for this example is 0.5944. The next step is to obtain the ratio, di/SD (see the fourth column of Table 3.2. In the fifth and sixth columns, respectively, the parameter n * erfc(abs(di)/SD is calculated and Chauvenet's test applied. Note that 'erfc' is the complementary error function and is available within MS Excel as an in-built function. One may access all in-built function within MS Excel/VBA via the code Application.erfc(z), where z is the parameter for which the complementary error function is required.

Discussion
We see that the third and fifth readings are suspect and hence the recommendation is to 'REJECT' those readings. The file Ex03-01.xls provides all the steps listed above.

Example 3.2 Solution of a fourth order algebra equation
Problem definition
Gerolamo Cardano was a sixteenth-century mathematician who published his *Ars Magna* in 1545 (Branson 2013). He is considered to be the originator of an analytical technique for solving cubic and quartic algebraic equations. A quartic algebraic equation is the highest order equation that may have an analytical solution, there being no generalized solution for quintic equation, i.e. an equation of five unknowns.

More recently, in the year 1897, H S Hall and S R Knight, both of Cambridge University, England published their treatise on 'Higher Algebra' [H S Hall and S R Knight, Higher algebra, Macmillan, London, 1897]. Hall and Knight have re-introduced the work of Cardano and presented a fourth order equation (Eq. 3.1) and its solution,

$$Y = X^4 - 2X^3 - 5X^2 + 10X - 3 = 0 \tag{3.1}$$

The roots of this equation are,

$$\frac{3 \pm \sqrt{5}}{2} \text{ and } \frac{-1 \pm \sqrt{13}}{2}, \text{ or}$$

$$2.618033989, \quad 0.381966011, \quad 1.302775638, \quad -2.302775638$$

Use the Bisection method to obtain the roots of Eq. (3.1).
Solution
Solving a quartic algebra equation via Bisection method.

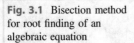

Fig. 3.1 Bisection method
for root finding of an
algebraic equation

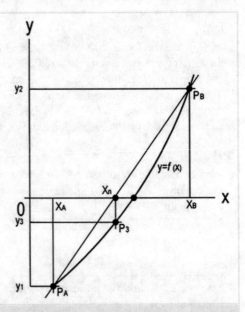

This method is also called the method of interval halving. It requires two initial approximations. The method will work if the initial guesses are on the opposite side of the root.

From Fig. 3.1, the two initial approximations are X_A and X_B and their mid-point is X_N. The function, Y is evaluated at the latter three points and then the following logic is used,

$$\text{If}(Y_1 * Y_3) > 0, \text{ then } X_A = X_n, \text{ else } X_B = X_n.$$

At each iteration Y_3 is obtained and the routine is halted provided Y_3 has reached a low value that is set by the allowable error tolerance. For the sake of safety a counter keeps track of the number of iterations and the routine terminated after a set limit is crossed.

The file Ex03-02.xlsm provides solution to this problem which is based on a macro. The VBA code for this macro is provided in Table 3.3.

Discussion

Tables 3.3 and 3.4, respectively, provide the VBA code for the Bisection method and the output generated in MS Excel. Note that this code will only provide roots of the given equation between the limits set in cells B15 and B16 (Table 3.4 refers). To obtain other roots those limits ought to be changed.

Table 3.3 VBA code for Bisection method

```
Sub BisectionMethod()
X1 = Sheets("Compute").Cells(15, 2)
X2 = Sheets("Compute").Cells(16, 2)
niter = 0
Line2:
X = X1
GoSub 101
Y1 = Y

X = X2
GoSub 101
Y2 = Y

XM = 0.5 * (X1 + X2)
X = XM
GoSub 101
YM = Y

' TOLERANCE FOR FUNCTION SET AT 1E-8
If (Abs(YM) < 0.00000001) Then
GoTo Line1
ElseIf (Y1 * YM) > 0 Then
X1 = XM
Else
X2 = XM
End If

If (niter > 1000) Then
GoTo Line1
Else
niter = niter + 1
Sheets("Compute").Cells(niter, 9) = X1
Sheets("Compute").Cells(niter, 10) = X2
Sheets("Compute").Cells(niter, 11) = YM
GoTo Line2
End If

Line1:
Sheets("Compute").Cells(18, 2) = XM
Sheets("Compute").Cells(18, 4) = niter
End

101 'Subroutine for function evaluation
Y = X ^ 4 - 2 * X ^ 3 - 5 * X ^ 2 + 10 * X - 3
Return
End Sub
```

Table 3.4 Output generated in MS Excel

	A	B	C	D	E	F
1	Ex 3.2 Solution of a quartic equation			X1	X2	FUNCTION
2	Bisection method			-3	-2	-11
3	Equation			-2.5	-2	11.0625
4	$X^4 - 2X^3 - 5X^2 + 10X - 3 = 0$			-2.5	-2.25	-2.40234375
5				-2.375	-2.25	3.65649414
6				-2.3125	-2.25	0.46705627
7				-2.3125	-2.28125	-1.01E+00
8				-2.3125	-2.296875	-0.27965444
9				-2.3046875	-2.296875	0.09121693
10				-2.3046875	-2.3007813	-0.09483775
11				-2.3046875	-2.3027344	-0.00196541
12				-2.3037109	-2.3027344	0.04458698
13				-2.3032227	-2.3027344	0.02130109
14				-2.3029785	-2.3027344	0.00966542
15	Lower bound	-3.00		-2.3028564	-2.30273438	0.0038494
16	Upper bound	-1.00		-2.3027954	-2.3027344	0.00094184
17				-2.3027954	-2.3027649	-0.00051182
18	**Root**	-2.3027756		-2.3027802	-2.3027649	0.000215
19	**Iterations**	31		-2.3027802	-2.3027725	-0.00014841
20				-2.3027763	-2.3027725	3.3293E-05
21				-2.3027763	-2.3027744	-5.7561E-05
22				-2.3027763	-2.3027754	-1.2134E-05
23				-2.3027759	-2.3027754	1.0579E-05
24				-2.3027759	-2.3027756	-7.7727E-07
25				-2.3027757	-2.3027756	4.9011E-06
26				-2.3027757	-2.3027756	2.0619E-06
27				-2.3027757	-2.3027756	6.4231E-07
28				-2.3027757	-2.3027756	-6.748E-08
29				-2.3027756	-2.3027756	2.8742E-07
30				-2.3027756	-2.3027756	1.0997E-07
31				-2.3027756	-2.3027756	2.1244E-08
32				-2.3027756	-2.3027756	-2.3118E-08

Example 3.3 Solution of a system of algebra equations: Hooke and Jeeves method

Problem definition

Amina and Fariha are two Afghan sibling princesses. Presently, Fariha is double the age of Amina but in 16 years' time the above ratio will become one-and-a-third. Find the respective ages of Amina and Fariha.

Let us denote the respective ages of Amina and Fariha as $X(1)$ and $X(2)$. Note that the above problem can be translated into two simultaneous equations (Eqs. 3.2 and 3.3),

$$X(2) = 2 * X(1) \tag{3.2}$$

$$X(2) + 16 = 1.3333 \, (X(1) + 16) \tag{3.3}$$

It may be possible to solve above equations via the matrix procedures introduced in Chaps. 1 and 2 but here we introduce a much more potent method that can not only handle a system of linear but also nonlinear equations. These procedures belong to that branch of mathematics that is known as 'optimisation'. Using optimization techniques you may solve simultaneous equations, fit linear or nonlinear models or indeed obtain the maximum and minimum of a given function. There are a great many applications of optimization within all branches of engineering.

Equations 3.2 and 3.3 may easily be converted into an optimization problem as shown in Eq. (3.4),

$$Z = [X(2) - 2 * X(1)]^2 + [\{X(2) + 16\} - \{1.3333 \, (X(1) + 16)\}]^2 \tag{3.4}$$

Or,

$$Z = [X(2) - 2 * X(1)]^2 + [X(2) - 1.3333 \, X(1) - 5.3328]^2 \tag{3.5}$$

Note that once $X(1)$ and $X(2)$ have converged, Z will assume nil value. We shall try to find those values of $X(1)$ and $X(2)$ that return a minimum value of Z.

Solution

There are a number of optimization methods that can easily handle the present task. We shall demonstrate a direct search method which is known as Pattern Search and is due to Hooke and Jeeves (R Hooke and T A Jeeves, Direct search solution of numerical and statistical problems. Journal of Association of Computing Machinery, 8, 1961).

Note that presently we are dealing with two independent variables and therefore the envelope of the dependent variable, Z will be described via three-dimensional geometry. If you have 'N' independent variables, then you are dealing with N + 1 dimensions.

In Pattern Search one makes two types of moves: exploratory and pattern moves. The exploratory move consists of a restricted search move, one variable at a time. The knowledge gained from the latter moves indicate whether, with each variable change, the dependent function Z reduces or not, the latter being the case for function minimization. The exploratory moves then provide enough information for a 'pattern' move, the latter being of a larger magnitude than the exploratory move.

Launch the Ex03-03.xlsm VBA program provided at this book's companion website. There is only one worksheet, 'Data' that contains the input and output data for the present example. The input data is keyed in the yellow-colored cells while the green-colored cells have the generated output. All variables have been suitably labeled.

INPUT DATA: N is the number of independent variables, Step length is related to the size of the move you select for pattern search and $\times$ 1, $\times$ 2, ...,xn are the initial values you provide.

OUTPUT DATA: The entries in column A are clear. The outputs in rows 10, 11, ... are the values of the independent variables (or solution) that are obtained.

The solution thus obtained is telling us that the present, respective ages of Amina and Fariha are 8 and 16 years

Discussion

In the above example a very powerful technique, known as optimization, was introduced. This technique may be used to solve a system of linear or non-linear algebra equations as well as finding the minimum or maximum of a given function. The latter usage is to be found in engineering to a large extent. The user may use the above code to solve problems of their own.

Example 3.4 Comparison of calculation of fluid-flow friction factor via an approximate and precise method

The present example considers the flow of any given fluid within a conduit, and the subsequent loss of energy due to the effects of fluid viscosity. When a fluid moves past a solid interface, a boundary layer is formed adjacent to the interface. The viscosity property of the fluid dictates a variation of velocity normal to the interface from a nil value at the wall (adhesion) to a maximum value at the outer edge of the boundary layer. This variation in the fluid velocity represents a loss in momentum and hence a resistance to the flow (drag force).

Problem definition

Fluid-flow pressure loss (Δp) may be shown to be directly related to the length (L) of the conduit, the fluid density (ρ), the square of the fluid velocity

(V), and inversely related to the conduit diameter (d). Equation 3.6 shows this relationship,

$$\Delta p = fr(L/d)\ (V^2/2) \tag{3.6}$$

The variable 'f' in the above equation represents the Darcy-Weisbach friction factor. Consider the flow of air under standard conditions, through a 4.0 mm diameter drawn tubing (roughness, $\varepsilon = 0.0015$ mm) with an average velocity of 50 m/s. Determine the pressure drop in a 0.1 m section of the tube.
Solution 1, using an approximate method that provides one-step answer

- Open the workbook Ex03-04a.xls. This workbook consists of one sheet, namely 'Compute' which includes given data as well as the computations performed by Excel.
- Read the example text carefully and insert the above given data in cell range (**B4:B7**). Under standard temperature and pressure, the density of air (ρ) = 1.23 kg/m^3, and its viscosity (μ) = 1.79 × 10^{-5} Ns/m^2. Insert these values in (**cell B8**) and (**cell B9**), respectively.
- The solution is shown in (**cell B18**). Excel performs all calculations as given below.
  - The Reynolds number, $Re = \frac{\rho V D}{\mu} = 1.37 \times 10^4$ (**cell B15**). Thus, the flow is turbulent (Re > 3000).
    The relative roughness, $\frac{\varepsilon}{D} = 3.75 \times 10^{-4}$ (**cell B16**).
  - Using the Swamee and Jain equation (Swamee and Jain 1976), the friction factor is obtained:
    $f = \dfrac{1.325}{\left[\ln\left(\frac{\varepsilon/D}{3.7}\right) + \frac{5.74}{Re^{0.9}}\right]^2} = 0.0292$ (**cell B17**). Note that the friction factor can also be found by using other models such as Colebrook equation (Colebrook 1938–39). In this respect see Example 2.9.1 or refer to Moody chart (Moody 1944) which gives (f = 0.0298).
    Note that the Swamee-Jain Equation has an accuracy of 99% when compared to Colebrook equation for: 10^{-6} < relative roughness < 10^{-2} and 5,000 < Re < 10^8.
  - The pressure drop, $\Delta p = f\frac{l}{D}\frac{1}{2}\rho V^2 = 1122$ Pa (**cell B18**). Note that you can explore the effect of the pipe diameter, length, roughness, or the flow velocity on the pressure drop by changing the input data in cell range (**B4:B9**).

Solution 2 using precise method that requires an iterative approach
This procedure may be used to obtain solution of any nonlinear algebra equation by Newton's method (Goal Seek).
Design engineers often encounter a situation where the result of a formula, or the final result at the end of a series of computations is known (or desired),

but not the input value. To solve such a problem the Excel **Goal Seek** facility may be used. Excel varies the value in the specified cell until the result in the target cell is obtained. The present example will be a good demonstration.

Newton's method

The *Newton,* or *Newton–Raphson,* method of approximating roots is probably the most widely used method because of its rapid convergence and ease of programming. It requires only one first guess which, if not sufficiently close to the root, may sometimes cause divergence, or convergence on the wrong root. To use this method it is necessary to know the first derivative of the given function.

Referring to Fig. 3.2, we make the initial guess X_B, from which we compute the ordinate Y_B. Evaluating the first derivative of our function at X_B, we obtain the slope m of the tangent to the curve at P_B, which crosses the x axis at X_n. Refer to Eqs. 3.7 and 3.8. We take X_n as our second approximation to the root. Since

$$f'(X_B) = \frac{(Y_B - 0)}{(X_B - X_n)} \tag{3.7}$$

we get,

$$X_n = X_B - \frac{Y_B}{f'(X_B)} \tag{3.8}$$

Our new approximation X_n is now used to compute Y_n and the slope at P_n from which a third approximation X_A is found, and so on until the required accuracy is achieved.

Excel's **Goal Seek** function is based on the Newton–Raphson method, and this is demonstrated by the following example.

Fig. 3.2 The Newton–Raphson method for solving nonlinear algebraic equations

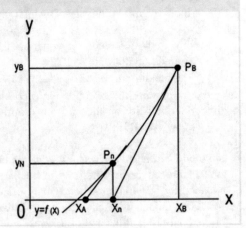

Colebrook's (1938–39) is the most precise model to obtain the fluid friction factor. It is represented by Eq. (3.9),

$$\frac{1}{\sqrt{f}} = -2\log_{10}\left(\frac{\varepsilon/D}{3.7} + \frac{2.51}{Re\sqrt{f}}\right) \tag{3.9}$$

Solution

- Open the workbook Ex03-04b.xls.
  - Colebrook's nonlinear equation requires a trial and error solution for friction factor. Therefore, assume a value for 'f' and insert it in **(cell B17)**. Convert Colebrook's equation to the form F(f) = 0 and insert it in **(cell B18)**. Use the Goal Seek function to find the value of 'f' **(cell B17)** that makes the value of **(cell B18)** = 0. Goal Seek is based on Newton's method and this works by starting with an initial value in the target cell, then zeroing in on a solution (f = 0.0291).
  - The pressure drop, $\Delta p = f \frac{L}{D}\frac{1}{2}\rho V^2 = 1119$ Pa **(cell B19)**, is in good agreement with the above obtained value of 1122 Pa

The use of the Goal Seek procedure is now explained in a step-by-step manner.

a. From the **Tools** menu select **Goal Seek**. If for any reason the **Goal Seek** function was not originally installed with Excel software then run the set-up program to install it.
b. In the **Set cell** box, enter the name of the target (or results) cell. In this particular instance the cell address is B18 and its value is 4.57. This quantity corresponds to the assumed value of friction factor, f = 0.01.
c. In the **To value** box, enter the value of zero.
d. In the **By changing cell** box, enter the reference of the cell which contains the value for the friction factor, i.e. B17.

Click on the **OK** button to initiate the Goal Seek process. If Goal Seek is not able to find a solution, it displays an error message. If it finds a solution, it shows the **Goal Seek Status** box which gives the **Current Value** to which the solution converges, 6.39×10^{-6} in the present case. If this value is acceptable, click **OK**.

Figure 3.3 shows an image of the **Goal Seek** dialog box with the cells correctly chosen.

Fig. 3.3 The Goal Seek
dialog box

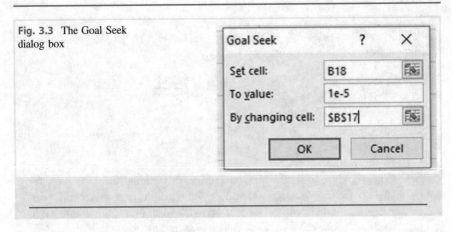

3.2 Conclusion

In this chapter, we have presented examples related to quality control of data using known and established statistical procedures. Optimization procedures were also demonstrated that enable solution of linear and nonlinear algebraic equations. The examples presented will find ready applications within engineering design and practice.

Exercises

E3.1 Chauvenet's criterion was introduced via Example 3.1. Develop an Excel/VBA code to enable construction of Table 3.2.

E3.2 Colebrook's model for obtaining the friction factor related to fluid flow in pipes and ducts is given by Eq. 3.4.4,

$$\frac{1}{\sqrt{f}} = -2\log_{10}\left(\frac{\varepsilon/D}{3.7} + \frac{2.51}{Re\sqrt{f}}\right)$$

Use the Bisection method for solving this nonlinear equation for the data given in Example 3.4.

E3.3 Aishah is a Canton-based chemist who wishes to know the minimum solubility, S of a compound that she has created in her laboratory. The relationship of the solubility against temperature, t is:

$$S(t) = t^4 - 12\ t^4 + 15t^4 + 56t - 60$$

This minimization exercise may easily be converted to a root finding problem by differentiation of the function and then obtaining the root. Use the bisection method to obtain the root within the interval $3 < t < 9$.

Refer to Example 3.2 for algorithmic details of the bisection method.

Answer: $t = 7.81$, $S = -703.73$

E3.4 Refer to Example 3.4 wherein the Newton's method of root finding was presented. Redo Exercise 3.3 by replacing bisection method with Newton's method.

E3.5 Rafay of Leithville is a process engineer whose laboratory tests have provided the following data that relate the saturation pressure, p_{sat} and saturation temperature T:

T, K	p_{sat}, kPa
330	1.208
340	1.667
350	2.251
360	2.983
370	3.883
380	4.978
390	6.290
400	7.846
410	9.674
420	11.923

Rafay wishes to obtain a relationship of the form: $p_{sat} = \text{Exp}[A - B/(T + C)]$. Using the MS Excel Solver facility, find the optimum values of A, B, and C. Hint: The above exercise may easily be converted to a function minimization problem by creating a table such as the one given below:

T, K	p_{sat}, kPa	P_{model}	Residual
330	1.208		
340	1.667		
350	2.251		
360	2.983		
370	3.883		
380	4.978		
390	6.290		
400	7.846		
410	9.674		
420	11.923		
		Sum	

The third column with the heading 'p_{model}' may be generated using trial values of A, B, and C. The fourth column with the heading 'Residual' may then be obtained as = Abs($p_{sat} - p_{model}$) and then the Sum will have the total of the Residual values. The Sum will thus become the Objective Function which needs to be minimized using MS Excel Solver.

Answer: A = 10.983, B = 3648.5, C = 8.9.

E3.6 MoGulam is an industrial engineer who has obtained the following relationship for inventory cost of engine components:

$$C(x) = \frac{225,000}{x} + \frac{7x}{200} + 5500$$

At any given time, a minimum of 3000 components must be held in stock. Find that value of x that minimizes C(x).

Note: the above exercise is constrained optimization problem with the constraint being x > 1000. You may use MS Excel Solver for this task.

Note that you need to provide the constraint within the Solver facility shown in Fig. 10.2. You need to use the 'Subject to constraints:' dialog box and insert the constraint using the 'Add' button.

References

W.B. Branson,Solving the Cubic with Cardano. Convergence (September 2013). http://www.maa.org/press/periodicals/convergence/solving-the-cubic-with-cardano.

C.F. Colebrook, Turbulent flows in pipes, with particular reference to the transition region between the smooth and rough pipe laws. J. Inst. Civ. Eng., London, 11, 133–156 (1938–39)

L.F. Moody, Friction factors for pipe flow. Trans ASME 66(8), 671–684 (1944)

P.K. Swamee, A.K. Jain, Explicit equation for pipe-flow problems. Proc, of the ASCE J. of the Hydraulics Division 102(HY5), 657–665 (1976)

Computer Graphics in Excel

4

4.1 Computer Graphics with MS Excel and VBA

MS Office products offer many options for manually or programmatically generating graphics. Unfortunately, they are not popular enough. Most people perceive
MS Excel as a spreadsheet and calculation program, and the maximum they can
imagine as possibilities is to plot a chart on that data.

In fact, MS Excel allows us to work with raster graphics and vector objects
(shapes), such as line segments, arrows, rectangles, triangles, polygons, polylines,
and more. This addition creates many possibilities, especially in combination with
VBA.

The following examples show 'tricky' approaches with which MS Excel can
perform some unexpected graphical tasks. These approaches can be classified as
follows:

- Pseudo-raster graphics—by colored cells (type 1);
- Pseudo-raster graphics—through Shapes collection (type 2);
- Vector graphics;
- Combination of raster (bitmap) and vector graphics;
- Visualization of 3D objects.

Example 4.1.1 RGB definition and use of gray shades
Problem definition
It is widely known that MS Excel is able to display data in shades of gray and
colors. When programmatically controlled, the *RGB*() function can be used,

---

Supplementary Information The online version contains supplementary material available at
https://doi.org/10.1007/978-3-030-94085-0_4.

T. Muneer and S. Ivanova, *Excel-VBA*,
https://doi.org/10.1007/978-3-030-94085-0_4

Table 4.1 Function *GrayColor*() and procedure *Draw_GrayShades*(), saved in Module1 in Ex04-01.xlsm

```
Function GrayColor(ByVal Value#, ByVal Min#, ByVal Max#) As Integer
'gray color adjustment function for visualization
    GrayColor = (Value - Min) / (Max - Min) * 255
End Function

Sub Draw_GrayShades()
Dim i%, Min%, Max%, gray%
    Min = 1: Max = 16
    For i = Min To Max
            gray = GrayColor(i, Min, Max)
            ActiveSheet.Cells(1, i).Interior.Color = RGB(gray, gray, gray)
    Next
End Sub
```

which receives three arguments for the primary colors red, green, and blue with values between 0 and 255. If all three components are the same in value, gray color is displayed. If this value is closer to 0, the resulting color is getting darker gray and vice versa—if they are closer to 255, the color is getting closer to white.

Let's start with a creation of a function that, by a given value and the left and right limits of the range in which this value changes, returns a number between 0 and 255, which is then used as an argument to the *RGB*() function. When the value coincides with its minimum possible value, the black color will be generated if it coincides with the maximum value—white color; if it is between the two borders—medium gray color.

Solution

The solution to this problem is demonstrated in the file Ex04-01.xlsm. Module1 includes a function *GrayColor*() and a executable sub-routine *Draw_GrayShades*(), see Table 4.1.

- The function *GrayColor*() returns a value between 0 and 255. The arguments are *Value* and lower limit *Min* and upper limit *Max*, which define the range of *Value*.
- The sub-routine *Draw_GrayShades*() is easy to start from the list of available macros or with a button. It uses the property. *Interior.Color* and function *RGB*() to change the background color of cells to a gray shade.

Discussion

As a result of the executed procedure, 16 cells of row 1 in the active sheet are colored in suitable shades of gray (Fig. 4.1) corresponding to the position of the value *i* between Min = 1 and Max = 16. Passing the same value for the three components to the RGB() function generates a gray

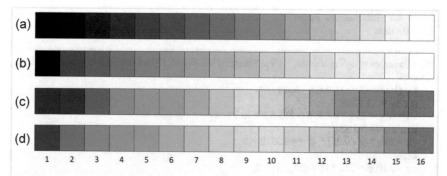

Fig. 4.1 Gray and color shades for values between Min = 1 and Max = 16: **a** gray without gamma correction; **b** gamma-corrected gray colors; **c** false colors without gamma correction; **d** gamma-corrected false colors. The image is generated within Ex04-03.xlsm

shade. RGB(0,0,0) generates a code for black color for $i = 1$ and RGB (255,255,255)—for white color for $i = 16$, and for values of the variable *gray* between 0 and 255, we get gray shades.

Example 4.1.2 RGB definition and use of false- and pseudo-colors
Problem definition
Color graphics are much more attractive than those in shades of gray. Sometimes it is necessary to show a grayscale image in color in order to better perceive the information in it by viewers. If it corresponds to values that cannot be perceived by the human eye due to its limitations to distinguish colors only with a wavelength of 300–700 nm, these colors are called *false colors*. If an image in grayscale (representing, for example, temperatures, or albedo, or atmospheric pressure, etc.) is to be depicted in color, it is called a *pseudo-color* image. In both cases, it is important where the value is in the range between the lowest and highest possible. If it is close to the lowest possible, it is usually colored dark blue and vice versa—if it is close to the highest possible, it is colored red (or magenta).

Create a procedure that, by a given value and the left and right limits (minimum and maximum) of the range in which this value changes, returns three numbers for the three primary colors, red, green, and blue components between 0 and 255, which are then used as arguments to the *RGB()* function.
Solution
The solution to this problem is demonstrated in the file Ex04-02.xlsm. Module1 includes a sub-routine *FalseColor()* and a executable sub-routine *Draw_FalseColorShades()*, see Table 4.2.

Table 4.2 Sub-routines *FalseColor*() and *Draw_FalseColorShades*(), saved in Module1 in Ex04-02.xlsm

```
Sub FalseColor(ByVal Value#, ByVal Min#, ByVal Max#, ByRef r%, ByRef g%, ByRef b%)
Dim c#
    c = (Value - Min) / (Max - Min)
    If c <= 0.25 Then
      r = 0: g = 255 * c * 4: b = 255 * 2 * (0.25 - c)
    ElseIf c <= 0.5 Then
      r = 255 * (c - 0.25) * 4: g = 255: b = 0
    Else
      r = 255: g = 255 * (1 - c) * 2: b = 0
    End If
End Sub

Sub Draw_FalseColorShades()
Dim i%, Min%, Max%, r%, g%, b%
    Min = 1: Max = 16
    For i = Min To Max
            Call FalseColor(i, Min, Max, r, g, b)
            ActiveSheet.Cells(1, i).Interior.Color = RGB(r, g, b)
    Next
End Sub
```

- The sub-routine *FalseColor*() determines the color's RGB components that correspond to the variable *Value* passed as an argument, depending on the difference between it and the predetermined limits Min and Max (Fig. 4.1). RGB components are returned as three separate integer arguments (between 0 and 255), submitted by address (with *ByRef*). Because such submitting (by address) is default in the VBA language, the *ByRef* keyword may be omitted. The procedure needs to return not one value (as in the *GrayColor*() function) but three values (*r, g,* and *b*); thus, it is better to implement it as a Sub.
- The main executable sub-routine *Draw_FalseColorShades*() uses the property.*Interior.Color* and function RGB() with three different arguments *r, g, and b* to change the background color of cells.

Discussion

As a result of executed procedure *Draw_FalseColorShades*(), 16 cells of row 1 in the active sheet are colored in suitable color shades corresponding to the position of the value i between Min = 1 and Max = 16 (from dark blue for *i* = 1 to red for *i* = 16 – Fig. 4.1).

The coordinate system of Excel cells is *right-oriented*, the X-axis points down, and the Y-axis points to the right.

Example 4.1.3 RGB definition and of false- and pseudo-colors with gamma correction

Problem definition

The human eye perceives light in a nonlinear way. Human vision is less sensitive to highlights than to dark shades. Thus, a *gamma* correction has to be applied to the computer colors (Fig. 4.1). For most current display systems, the encoding gamma value is ≈ 0.45, and the reciprocal decoding value is ≈ 2.2. The gamma value should be used as a power indicator of the numerical value of the color light intensity, which is in a range from 0 (for black) to 1 (for maximum intensity). After this operation, the result obtained is multiplied by 255 so that it can be passed to the *RGB()* function.

Edit the function *GrayColor()* and procedure *FalseColor()* so that a gamma correction is applied to the generated gray and color shades.

Table 4.3 Function GrayColor_GammaCorrected(), procedures FalseColor_GammaCorrected (), and Draw_FalseColorShades_gamma_corrected(), saved in Module1 in Ex04-03.xlsm

```
Public gamma_correction#

Function GrayColor_GammaCorrected(ByVal Value#, ByVal Min#, ByVal Max#) As Integer
'gray color adjustment for visualization with public variable gamma_correction=0.45
Dim g#
   g = (Value - Min) / (Max - Min)
   GrayColor_GammaCorrected = (g ^ gamma_correction) * 255
End Function

' color adjustment procedure for visualization
' set a public variable gamma_correction=0.45
Sub FalseColor_GammaCorrected(ByVal Value#, ByVal Min#, ByVal Max#, ByRef r%, ByRef g%, ByRef b%)
Dim c#, rr#, gg#, bb#
   c = (Value - Min) / (Max - Min)
   If c <= 0.25 Then
      rr = 0: gg = c * 4: bb = 2 * (0.25 - c)
   ElseIf c <= 0.5 Then
      rr = (c - 0.25) * 4: gg = 1: bb = 0
   Else
      rr = 1: gg = (1 - c) * 2: bb = 0
   End If
' gamma correction
   r = (rr ^ gamma_correction) * 255
   g = (gg ^ gamma_correction) * 255
   b = (bb ^ gamma_correction) * 255
End Sub

Sub Draw_FalseColorShades_gamma_corrected()
Dim i%, Min%, Max%, r%, g%, b%, gray%
   Min = 1: Max = 16
   gamma_correction = 0.45
   For i = Min To Max
         gray = GrayColor_GammaCorrected(i, Min, Max)
         ActiveSheet.Cells(3, i).Interior.Color = RGB(gray, gray, gray)
         Call FalseColor_GammaCorrected(i, Min, Max, r, g, b)
         ActiveSheet.Cells(7, i).Interior.Color = RGB(r, g, b)
   Next
End Sub
```

Solution

The solution to this problem is demonstrated in the file Ex04-03.xlsm. Module1 includes a function *GrayColor_GammaCorrected*(), a sub-routine *FalseColorShades_gamma_corrected*(), and a executable sub-routine *Draw_FalseColorShades_gamma_corrected*(), see Table 4.3.

- The variable *gamma_correction* is defined as *Public* at the top of Module1. It is very important that this variable is set to 0.45 before its first use.
- The function *GrayColor_gamma_corrected*() has been modified in comparison with the function *GrayColor*() in Example 4.1.1 to prepare an adjusted gray color gamut.
- The sub-routine *FalseColorShades_gamma_corrected*() is based on the sub-routine *FalseColor*() in Example 4.1.2 to prepare an adjusted color gamut.
- The sub-routine *Draw_FalseColorShades_gamma_corrected*() sets the variable *gamma_correction* to 0.45 and calls other sub-routines.

Discussion

The cells in rows 1 and 5 are already colored in non-gamma-corrected colors for a comparison (Fig. 4.1). As a result of the last procedure, 16 cells in row 3 in the active sheet are colored in gamma-corrected gray shades (Fig. 4.1), and 16 cells in row 7 in the active sheet are colored in gamma-corrected colors (Fig. 4.1).

Example 4.1.4 Pseudo-raster graphics, using colored Excel cells

Problem definition

The cells in the MS Excel worksheet could be used as pixels in pseudo-raster graphics. Each cell can be made square and colored in the desired color (without text inside); thus, it forms a 'pixel'. In this way, we can generate both grayscale and full-color graphics.

Generate a false-colored rectangular graphic that reflects the contents of a two-dimensional array (matrix) of numeric values (Fig. 4.2).

Solution

The solution to this problem is demonstrated in the file Ex04-04.xlsm. Module1 includes already mentioned sub-routine *FalseColor_GammaCorrected*(), two new functions *MinMatrix*() and *MaxMatrix*(), and two sub-routines *DrawMatrix()* and *Main*().

- The function *MinMatrix*() returns the lowest value of the array Matrix() with M rows and N columns.

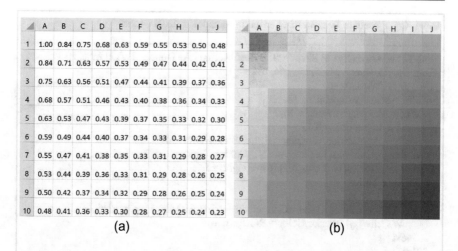

	A	B	C	D	E	F	G	H	I	J
1	1.00	0.84	0.75	0.68	0.63	0.59	0.55	0.53	0.50	0.48
2	0.84	0.71	0.63	0.57	0.53	0.49	0.47	0.44	0.42	0.41
3	0.75	0.63	0.56	0.51	0.47	0.44	0.41	0.39	0.37	0.36
4	0.68	0.57	0.51	0.46	0.43	0.40	0.38	0.36	0.34	0.33
5	0.63	0.53	0.47	0.43	0.39	0.37	0.35	0.33	0.32	0.30
6	0.59	0.49	0.44	0.40	0.37	0.34	0.33	0.31	0.29	0.28
7	0.55	0.47	0.41	0.38	0.35	0.33	0.31	0.29	0.28	0.27
8	0.53	0.44	0.39	0.36	0.33	0.31	0.29	0.28	0.26	0.25
9	0.50	0.42	0.37	0.34	0.32	0.29	0.28	0.26	0.25	0.24
10	0.48	0.41	0.36	0.33	0.30	0.28	0.27	0.25	0.24	0.23

(a)   (b)

Fig. 4.2 Visualization of matrix (10×10) with numeric values: **a** matrix; **b** visualized matrix. The minimum is 0.23 (in blue); the maximum is 1 (in red). The image is generated within Ex04-04.xlsm

- The function *MaxMatrix*() returns the highest value of the array Matrix() with M rows and N columns.
- The sub-routine *DrawMatrix*() visualizes the data matrix in false colors. It receives as arguments the sheet name where the visualization will be the data matrix itself and its number of rows and columns, see Table 4.4. It displays the lowest value Min in dark blue, and the largest Max in red (as shown in Fig. 4.1).
- The sub-routine *Main*() calls other sub-routines.

Discussion

A vital element of the procedure *DrawMatrix*() is adjusting the width and height of the cells to form a square. Two lines in the VBA code change the width of *N* columns to 0.5 characters and the height of *M* rows in the sheet to 5 points. Although the column width and row height units are different, visually, 0.5 characters (for font size 10) are equal to 5 points, and each cell appears on the screen as a small square pixel. All cells form a raster image, which will be further colored.

The next step is to change the background color of $M \times N$ empty cells in the sheet. The property.*Interior.Color* does this in combination with the *RGB* () function, which receives three arguments of the color components with values between 0 and 255.

This procedure can visualize the values in any matrices. Here are some possible applications:

- Visualization of the parameters' values of a studied surface, determined using the finite element method;

Table 4.4 Procedure *DrawMatrix*(), saved in Module1 in Ex04-04.xlsm

```
Sub DrawMatrix(ByVal SheetStr$, ByRef Matrix() As Double, ByVal M%, ByVal N%)
Dim i%, j%, Max#, Min#, r%, g%, b%
Dim W As Worksheet
   Min = MinMatrix(Matrix, M, N)
   Max = MaxMatrix(Matrix, M, N)
   gamma_correction = 0.45
   Set W = Application.Worksheets(SheetStr): W.Activate
   W.Range(Columns(1), Columns(N)).ColumnWidth = 0.5
   W.Range(Rows(1), Rows(M)).RowHeight = 5
   For i = 1 To M
    For j = 1 To N
       Call FalseColor_GammaCorrected(Matrix(i, j), Min, Max, r, g, b)
       W.Cells(i, j).Interior.Color = RGB(r, g, b)
    Next
   Next
End Sub
```

- Visualization of national, continental, and world maps of various parameters of land or water surface (temperature, radiation, albedo, atmospheric pressure, etc.).

Example 4.1.5 Visualization of NASA maps in Excel cells
Problem definition
Diverse, useful, and interesting data for the second option can be downloaded from the NASA website (NEO—NASA Earth Observation website) with different resolution: 1.0 degree (matrix size 180×360), 0.5 degrees (matrix size 360×720), 0.25 degrees (matrix size 720×1440), and 0.1 degrees (matrix size 1800×3600).

Some of these NASA's data matrices include values that should not be visualized. For example, if the matrix contains values of the average daily temperatures of the land surface, the pixels that correspond to oceans and seas should be colored in black. Therefore, the fixed value 99,999 is stored in the corresponding elements in the matrices. This fact requires a modified function *MaxMatrix_NASA*() for searching for a maximum value that ignores the value 99,999 as the highest possible value (see Table 4.5).

Write the necessary procedures and functions with which to draw graphics on various datasets downloaded from the NASA website with a resolution of 1 degree (NEO—NASA Earth Observation website).

Table 4.5 Function *MaxMatrix_NASA*() and procedure *DrawMap*(), saved in Module1 in Ex04-05.xlsm

```
Function MaxMatrix_NASA(ByRef X() As Double, ByVal M%, ByVal N%) As Double
Dim i%, j%, Max#
    Max = -1000000
    For i = 1 To M
      For j = 1 To N
        If Max <X(i, j) And X(i, j) <> 99999 Then Max = X(i, j)
      Next
    Next
    MaxMatrix_NASA = Max
End Function

Sub DrawMap(ByVal SheetStr$, ByRef Matrix() As Double, ByVal M%, ByVal N%)
'Visualization of matrix data in false colors

Dim i%, j%, Max#, Min#, r%, g%, b%
Dim W As Worksheet
    Max = MaxMatrix_NASA(Matrix, M, N)
    Min = MinMatrix(Matrix, M, N)
    Set W = Application.Worksheets(SheetStr): W.Activate
    W.Range(Columns(1), Columns(N)).ColumnWidth = 0.5
    W.Range(Rows(1), Rows(M)).RowHeight = 5

    For i = 1 To M
     For j = 1 To N
      If Matrix(i, j) <> 99999 Then
        Call FalseColor_GammaCorrected(Matrix(i, j), Min, Max, r, g, b)
        W.Cells(i, j).Interior.Color = RGB(r, g, b)
      Else
        W.Cells(i, j).Interior.Color = RGB(0, 0, 0) ' black
      End If
     Next
    Next
End Sub
```

Solution
The solution to this problem is demonstrated in the file Ex04-05.xlsm. Module1 includes already mentioned sub-routine *FalseColor_GammaCorrected*(), two functions *MinMatrix*() and *MaxMatrix_NASA*(), and two sub-routines *DrawMap()* and *Main*().

- The function *MaxMatrix_NASA*() returns the highest value of the array Matrix() with M rows and N columns.
- *DrawMap*() is a procedure similar to *DrawMatrix*() procedure in Example 4.1.4, but with some different program lines because it needs to be adapted for NASA's datasets. Pixels corresponding to map positions for which no

data is available must be displayed in black (i.e. outside the false-color range from dark blue to red, see Fig. 4.1), calling the *RGB* function with three zero arguments (0,0,0).

- The procedure *ConvertTo_TextFile*() processes the original CSV file, downloaded from the NASA site, and converts it to a normal text file.
- The procedure *ReadMatrix*() reads the map matrix with 180 lines and 360 columns from the text file.
- The procedure *DisplayMatrix*() writes the numeric values of the map matrix in a specified sheet.
- The sub-routine *Main*() calls other sub-routines.

Discussion

The numeric values of the data files downloaded from the NASA website in CSV format are separated by commas, and the individual text lines are separated by Line Feed character (LF) only, instead of the usual LF + CR combination. This makes it difficult to read the numeric matrix values from the files with the VBA command Input #. Therefore, a procedure named *ConvertTo_TextFile*() has been added to the program code that converts the CSV file into a plain text file with LF + CR at the end of each line. Reading the values of the matrix from the new text file is done with the procedure *ReadMatrix*().

The above VBA code in the MS Excel environment generates the images in Figs. 4.3, 4.4 and 4.5 using public data on the NEO, NASA website, with a resolution of 1 angular degree. Thus, the size of the generated raster images is 180 × 360 pixels (cells). The three CSV data files for these three figures, downloaded from the NASA site, are stored in a subfolder Files-for-Ex04-05.

Fig. 4.3 Example for pseudo-raster graphic, generated in Excel, for daily average land surface temperature for December 2001. *Data source* NASA (NEO—NASA Earth Observation website, average land surface temperature [day]). The image is generated within Ex04-05.xlsm

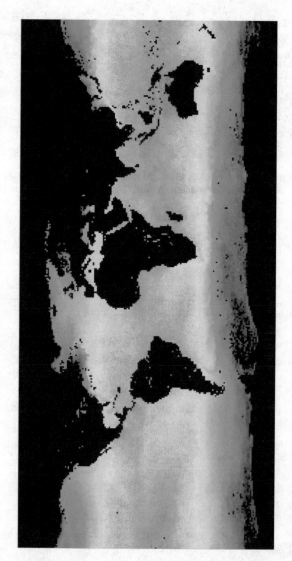

Fig. 4.4 Example for pseudo-raster graphic, generated in Excel, for average sea surface temperature for January 2003. *Data source* NASA (NEO—NASA Earth Observation website, average sea surface temperature). The image is generated within Ex04-05.xlsm

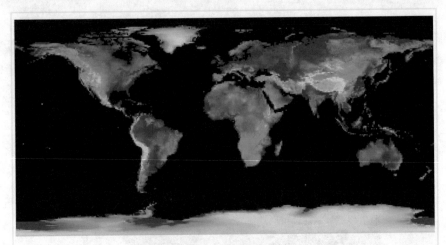

Fig. 4.5 Example for pseudo-raster graphic, generated in Excel, for topography for 2000. *Data source* NASA (NEO—NASA Earth Observation website, Topography). The image is generated within Ex04-05.xlsm

Example 4.1.6 Visualization of parts of NASA world maps in Excel cells

Problem definition

As mentioned above, higher resolution data—up to 0.1 angular degrees—can be downloaded from NASA's website. This option allows detailed maps to be plotted, for example, continental and national, using fewer data organized in a submatrix that is a part of the large downloaded matrix.

Write the necessary procedures and functions to draw graphics on a part of a dataset downloaded from the NASA website with a resolution of 0.1 degrees (Fig. 4.6) (NEO—NASA Earth Observation website). The arguments should include the latitudes and longitudes of the two opposite angles of the detailed map.

Solution

The solution to this problem is demonstrated in the file Ex04-06.xlsm. Module1 includes already mentioned sub-routine *FalseColor_GammaCorrected()*, *ConvertTo_TextFile()*, *ReadMatrix()*, and some new procedures (see Table 4.6):

- The function *MaxSubMatrix()* estimates the highest value of the array Matrix() between rows M1 and M2 and columns N1 to N2.
- The function *MinSubMatrix()* estimates the lowest value of the array Matrix() between rows M1 and M2 and columns N1 to N2.

Fig. 4.6 Example for pseudo-raster graphic, generated in Excel, for the daily average land surface temperature of Europe—May 2021, from Long: 10°, Lat: 35° to Long: 55°, Lat: 70°. *Data source* NASA (NEO—NASA Earth Observation website, daily average land surface temperature [day]). The image is generated within Ex04-06.xlsm

- *DrawSubMap*() is a procedure similar to *DrawMap*() procedure in Example 4.1.5. The arguments for the procedure are: the sheet name (where the visualization should take place); the global data matrix itself (size 1800 × 3600 elements), downloaded from the NASA website; longitude *Long1* and latitude *Lat1* of the bottom-left corner of the map; and longitude *Long2* and latitude *Lat2* latitude of the upper right corner of the map.
- The sub-routine *Main*() calls other sub-routines.

Discussion

The above-mentioned procedure *ConvertTo_TextFile*() in Example 4.1.5, which converts the CSV file to a plain text file, has also been added to this program code, as the *ReadMatrix*() procedure for reading the matrix values from the new generated text file.

The procedure *DrawSubMap*() requires new different minimum and maximum functions that search only part of a matrix, from *Row1* (corresponding to *Lat2*) to *Row2* (corresponding to *Lat1*), and from *Column1* (corresponding to *Long1*) to *Column2* (corresponding to *Long2*). The maximum finding function *MaxSubMatrix*() takes into account that the matrix has values equal to 99,999, which do not need to be displayed in color. The sheet Europe must be pre-created in advance so that the map can be drawn in it.

Table 4.6 Procedure *DrawSubMap*(), saved in Module1 in Ex04-06.xlsm

```
Sub DrawSubMap(SheetStr$, Matrix() As Double, Long1#, Lat1#, Long2#, Lat2#)
Dim i%, j%, Max#, Min#, step#, r%, g%, b%, Row1%, Column1%, Row2%, Column2%
Dim W As Worksheet, T%
  step = 0.1 ' resolution
  Row1 = (90 - step / 2 - Lat2) / step + 1
  Column1 = (Long1 + 180 - step / 2) / step + 1
  Row2 = (90 - step / 2 - Lat1) / step + 1
  Column2 = (Long2 + 180 - step / 2) / step + 1
  If Row1 > Row2 Then T = Row1: Row1 = Row2: Row2 = T
  If Column1 > Column2 Then T = Column1: Column1 = Column2: Column2 = T
  Max = MaxSubMatrix(Matrix, Row1, Column1, Row2, Column2)
  Min = MinSubMatrix(Matrix, Row1, Column1, Row2, Column2)
  Set W = Application.Worksheets(SheetStr): W.Activate
  W.Range(Columns(1), Columns(Column2 - Column1 + 1)).ColumnWidth = 0.5
  W.Range(Rows(1), Rows(Row2 - Row1 + 1)).RowHeight = 5

  For i = Row1 To Row2
    For j = Column1 To Column2
      If Matrix(i, j) <> 99999 Then
        Call FalseColor_GammaCorrected(Matrix(i, j), Min, Max, r, g, b)
        W.Cells(i - Row1 + 1, j - Column1 + 1).Interior.Color = RGB(r, g, b)
      Else
        W.Cells(i - Row1 + 1, j - Column1 + 1).Interior.Color = RGB(0, 0, 0)
      End If
    Next
  Next
End Sub
```

Figure 4.6 shows an example of daily averaged temperature map of Europe, displayed in the window from Long: 10°, Lat: 35° to Long: 55°, Lat: 70°. The CSV data file for the figure, downloaded from the NASA site, is stored in the subfolder Files-for-Ex04-06. The *DrawSubMap*() procedure could be called as follows:

Call DrawSubMap('Europe', Matrix, −10, 35, 55, 70)

Example 4.1.7 Pseudo-raster graphics using Shape collection
Problem definition

There is another possible solution to achieve pseudo-raster graphics in Excel —by using the graphic objects in the *Shape* collection, available in all products in MS Office. The list of different available shapes is accessible from tab Insert in MS Excel. The coordinate system in which the shapes are drawn is *left-oriented*, with the X-axis to the right and the Y-axis downwards, its

Fig. 4.7 Pseudo-color graphic of variations of the view factor between two surfaces with a common edge. According to legend, the minimum possible value is 0, the maximum—1. The angle between both surfaces is 45°. In this particular case, the minimum displayed value is 0.167, the maximum —0.784. The image is generated within Ex04-07.xlsm

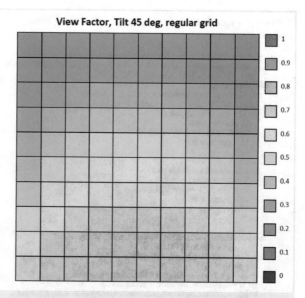

beginning is in the upper left corner of cell A1, and the dimensions are set in points (a point is 1/72 inch; font sizes are usually measured in points).

Generate a false-colored rectangular graphic that reflects the contents of a two-dimensional array (matrix) of numeric values (Fig. 4.7). Add to the graphic a legend about the numeric meaning of used colors.

Solution

The solution to this problem is demonstrated in the file Ex04-07.xlsm. Module1 includes already mentioned in the previous examples sub-routine *FalseColor_GammaCorrected()* and some new procedures (see Table 4.7):

- The sub-routine *Input_Matrix()* inputs the values of the matrix from the specified sheet.
- The sub-routine *Delete_AllShapes_Sheet()* removes all used shape objects in a specified sheet.
- *DrawRectangle()* is a procedure, drawing a rectangle on a specified sheet. It receives such arguments: *SheetStr* is the sheet name, $\times 1$ and *y1* are the coordinates of the upper left vertex of the rectangle, and *dx* and *dy*—horizontal and vertical size of the rectangle. *Rf, Gf,* and *Bf* are RGB components of its fill color, *Rl, Gl,* and *Bl* are RGB components of the frame line color of the rectangle, *Weight* is the thickness of the frame.
- The procedure *InsertText()* adds text to the sheet, and it receives next six arguments: sheet name *SheetStr*, *x* and *y* are coordinates where the text should be written, *txt* is the text, *Size* is the size of the text in points, *Bold* should be 0 to have normal text and 1—for bold.
- The procedure *Draw_View_Factor()* creates a rectangular graphic that displays the variations in the view factor between two surfaces with a

Table 4.7 Procedures *DrawRectangle*(), *InsertText*(), *Draw_View_Factor*(), saved in Module1 in Ex04-07.xlsm

```
Sub DrawRectangle(ByVal SheetStr$, ByVal x1#, ByVal y1#, ByVal dx#, ByVal dy#, _
        Rf%, Gf%, Bf%, Rl%, Gl%, Bl%, Weight%)
Dim W As Worksheet
  Set W = Application.Worksheets(SheetStr): W.Activate
  W.Shapes.AddShape(msoShapeRectangle, x1, y1, dx, dy).Select
  With Selection.ShapeRange.Fill
    .Visible = msoTrue
    .ForeColor.RGB = RGB(Rf, Gf, Bf)
    .Transparency = 0
    .Solid
  End With
  With Selection.ShapeRange.Line
    .Visible = msoTrue
    .ForeColor.RGB = RGB(Rl, Gl, Bl)
    .Transparency = 0
  End With
End Sub

Sub InsertText(ByVal SheetStr$, ByVal x#, ByVal y#, ByVal txt$, _
        ByVal Size%, ByVal Bold%)
Dim W As Worksheet, i%, j%, dx%, dy%
  dx = 50: dy = 20
  If Size <> 0 Then dx = Len(txt) * Size / 2: dy = Size * 1.5
  Set W = Application.Worksheets(SheetStr): W.Activate
  W.Shapes.AddTextbox(msoTextOrientationHorizontal, x, y, dx, dy).Select
  Selection.ShapeRange.Fill.Visible = msoFalse
  Selection.ShapeRange.Line.Visible = msoFalse
  Selection.ShapeRange(1).TextFrame2.TextRange.Characters.Text = txt
  If Bold = 1 Then
    Selection.ShapeRange.TextFrame2.TextRange.Font.Bold = msoTrue
  Else
    Selection.ShapeRange.TextFrame2.TextRange.Font.Bold = msoFalse
  End If
  If Size <> 0 Then Selection.ShapeRange.TextFrame2.TextRange.Font.Size = Size
End Sub

Sub Draw_View_Factor(ByVal SheetStr$, ByRef Matrix() As Double, _
  ByVal M%, ByVal N%, ByVal Min#, ByVal Max#, ByVal Size%, ByVal dxy_frame%)
Dim i%, j%, x0%, y0%, Size_Frame%
Dim r%, g%, b%, dy#, dyy#
  Size_Frame = Size + 2 * dxy_frame
  x0 = 10: y0 = 10

  Delete_AllShapes_Sheet(SheetStr)
  Call DrawRectangle(SheetStr, x0, y0, Size_Frame + 2 * dxy_frame, _
    Size_Frame, 255, 255, 255, 0, 0, 0, 1)
  For i = 1 To M
   For j = 1 To N
    Call FalseColor_GammaCorrected(Matrix(i, j), Min, Max, r, g, b)
    Call DrawRectangle(SheetStr, x0 + dxy_frame + (j - 1) * Size / N, _
      y0 + dxy_frame + (i - 1) * Size / M, Size / N, Size / M, _
      r, g, b, 0, 0, 0, 1)
   Next
  Next

  Call InsertText(SheetStr, x0 + 80, y0, "View Factor, Tilt 60 deg, regular grid", 16, 1)

  ' draw a legend
  x0 = x0 + Size + dxy_frame + 5 ' horisontal position of legend
  dy = Size / 21 * 2
  dyy = dy / 2

  For i = 1 To 11
    Call FalseColor_GammaCorrected((i - 1) / 10, Min, Max, r, g, b)
    Call DrawRectangle(SheetStr, x0 + 5, y0 + Size_Frame - dxy_frame - (i - 0.5) * dy, dyy, dyy, r, g, b, 0, 0, 0, 1)
    Call InsertText(SheetStr, x0 + 1.25 * dyy, y0 + Size_Frame - dxy_frame - (i - 0.5) * dy - dyy / 4, (i - 1) / 10, 0, 0)
  Next
End Sub
```

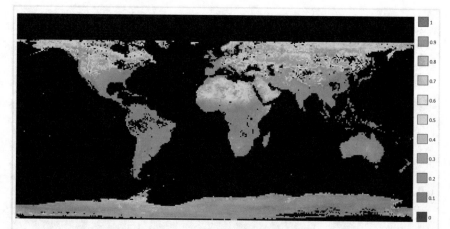

Fig. 4.8 Example for a combination of pseudo-raster graphic and shape objects, generated in Excel, for albedo map for January 2017. *Data source* NASA (NEO—NASA Earth Observation website, Albedo map). The image is generated within Ex04-08.xlsm

common edge. The procedure has some arguments: *SheetStr* is the sheet name, where the graphic will be arranged, *Matrix*() contains the dataset, M and N are numbers of rows and columns of the matrix, Min and Max are the bottom and upper limit of the values, *Size* in points is the size of the square colored graphic, *dxy_frame* is the size of the white field around the colored graphic.

• The sub-routine *Main*() calls other sub-routines.

Discussion

The procedure *Draw_View_Factor*() creates a rectangular graphic that displays the variations in the view factor between two surfaces with a common edge. How to calculate the values of VF in the different points of a surface is explained in (Muneer et al. 2015).

Legends, created with a code similar to above, could also be added to the pseudo-raster graphics described in the previous Examples 4.1.5 and 4.1.6 (Fig. 4.8 is generated within Ex04-08.xlsm).

Example 4.1.8 Combination of pseudo-raster graphics using cells and Shape collection

Problem definition

The maps generated in Examples 4.1.5 and 4.1.6 in Figs. 4.3, 4.4, 4.5 and 4.6 miss legends. Add a color legend to the maps.

Solution

The solution is demonstrated in the file Ex04-08.xlsm (see Fig. 4.8). Module1 includes a combination of procedures, already mentioned in Examples 4.1.5 and 4.1.7, and some new procedures:

- The procedure *DrawLegend*() has arguments: *SheetStr* is the sheet name, where the graphic will be arranged, *Min* and *Max* are the bottom and upper limit of the values.
- The procedure *DrawMapLegend*() draws the map, using sheet's cells and a legend, using shape rectangle and adds some text values to it.
- The sub-routine *Main*() calls other sub-routines.

Example 4.1.9 Drawing with vector primitives

Problem definition

The Shapes collection available in MS Excel is quite rich. Even the use of the simplest of them (line, circle/ellipse, arrow) in combination with VBA can help create quite complex graphics, not only in a plane but also axonometries and even perspectives. To begin with, let's look at a few procedures for drawing a line, circle, and arrow, which invoke not only the coordinates of the location of graphic elements but also their style (dashed or solid line and what kind of dashed line) and color. All figures in the following sub-sections of this chapter are generated in MS Excel using Shapes.

Create procedures to use those elements of the collection Shapes that are used to draw a circle, an ellipse, a regular polygon, and an arrow.

Solution

The solution is demonstrated in the file Ex04-09.xlsm. Module1 includes a combination of drawing procedures, and some of them are already mentioned —*DrawRectangle*(), *InsertText*(), *Delete_AllShapes_Sheet*()—see Table 4.8:

- The procedure *DrawLine*() has arguments: *SheetStr* is the sheet name, where the line will be displayed, $\times 1,y1$ and $\times 2,y2$ are the coordinates of the line's ends; the argument *Style* switches between two line styles in Excel; the argument *Dashed* draws dashed line for value 1 and continuous line for 0. Arguments *Rl, Gl, and Bl* are RGB values for the line color.
- The procedure *DrawCircle*() has arguments: *SheetStr* is the sheet name, x and y are the coordinates of the circle's center, and R is the radius.
- The procedure *DrawEllipse*() has arguments: *SheetStr* is the sheet name, x and y are the coordinates of the ellipse's center, and $R1$ and $R2$ are the semi-axes.
- The procedure *DrawArrow*() has arguments: *SheetStr* is the sheet name, 1 and $y1$ are the coordinates of the start point of the arrow, and $\times 2$, $y2$—the coordinates of the endpoint.

Table 4.8 Procedures *DrawLine()*, *DrawCircle()*, *DrawEllipse()*, *DrawArrow()*, *DrawPolygon ()*, *DrawPoint()*, saved in Module1 in Ex04-09.xlsm

```
Sub DrawLine(ByVal SheetStr$, ByVal x1#, ByVal y1#, ByVal x2#, ByVal y2#, _
   ByVal Style%, ByVal Dashed%, ByVal Rl%, ByVal Gl%, ByVal Bl%, ByVal Weight%)
Dim W As Worksheet
Set W = Application.Worksheets(SheetStr): W.Activate
   ActiveSheet.Shapes.AddConnector(msoConnectorStraight, x1, y1, x2, y2).Select
   If Style = 1 Then Selection.ShapeRange.ShapeStyle = msoLineStylePreset1
   With Selection.ShapeRange.Line
      If Dashed = 1 Then .DashStyle = msoLineDash
      .Visible = msoTrue
      .ForeColor.RGB = RGB(Rl, Gl, Bl)
      .Weight = Weight
      .Transparency = 0
   End With
End Sub

Sub DrawCircle(ByVal SheetStr$, ByVal x#, ByVal y#, ByVal R#)
Dim W As Worksheet
Set W = Application.Worksheets(SheetStr): W.Activate
   ActiveSheet.Shapes.AddShape(msoShapeOval, x - R, y - R, 2 * R, 2 * R).Select
   Selection.ShapeRange.ShapeStyle = msoShapeStylePreset1
End Sub

Sub DrawEllipse(ByVal SheetStr$, ByVal x#, ByVal y#, ByVal R1#, ByVal R2#)
Dim W As Worksheet
Set W = Application.Worksheets(SheetStr): W.Activate
   W.Shapes.AddShape(msoShapeOval, x - R1, y - R2, 2 * R1, 2 * R2).Select
   Selection.ShapeRange.ShapeStyle = msoShapeStylePreset1
End Sub

Sub DrawArrow(ByVal SheetStr$, ByVal x1#, ByVal y1#, ByVal x2#, ByVal y2#)
Dim W As Worksheet
Set W = Application.Worksheets(SheetStr): W.Activate
   ActiveSheet.Shapes.AddConnector(msoConnectorStraight, x1, y1, x2, y2).Select
      Selection.ShapeRange.Line.EndArrowheadStyle = msoArrowheadTriangle
End Sub

Sub DrawPolygon(ByVal SheetStr$, ByVal x#, ByVal y#, ByVal N%, ByVal Rcs#, ByVal Style%, ByVal Dashed%)
' Style: 1 - black, 0 - blue; Dashed: 0 - continuous line, 1 - dashed line
' Rcs - radius of circumscribed circle
Dim i%, A1#, A2#
   A1 = 0
   For i = 1 To N
      A2 = 2 * Application.Pi() / N * i
      Call DrawLine(SheetStr$, x + Rcs * Cos(A1), y - Rcs * Sin(A1), _
                    x + Rcs * Cos(A2), y - Rcs * Sin(A2), _
                    Style, Dashed, 0, 0, 0, 1)
      A1 = A2
   Next
End Sub

Sub DrawPoint(ByVal SheetStr$, ByVal x#, ByVal y#)
Dim W As Worksheet
Set W = Application.Worksheets(SheetStr): W.Activate
   ActiveSheet.Shapes.AddShape(msoShapeOval, x - 2, y - 2, 4, 4).Select
   Selection.ShapeRange.ShapeStyle = msoShapeStylePreset8
End Sub
```

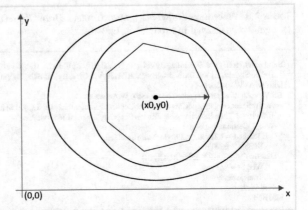

Fig. 4.9 Example for vector
graphics of XY coordinate
system: a line segment, a
circle, an ellipse, and a
regular polygon, generated in
Excel. The image is
generated within Ex04-09.
xlsm

- The procedure *DrawPolygon*() draws regular polygon. Its arguments are:
  SheetStr is the sheet name, *x and y* are the coordinates of the polygon's
  center, *N* is the number of the vertices of the polygon, *Rcs* is the radius of
  the circumscribed circle, the arguments *Style* and *Dashed* are the same, as
  in *DrawLine*().
- The procedure *DrawPoint*() draws a black circle as a point with radius 2,
  and *x and y* are the coordinates of the point's center.
- The procedure *Draw_Shapes*() calls the above sub-routines and creates the
  image in Fig. 4.9.

Discussion
The above procedures can be used to create more complex images with vector
graphic. Some possibilities are demonstrated in Sects. 4.2 and 4.3.

Example 4.1.10 Combining bitmap graphics with vector graphic
Problem definition
A helpful opportunity is to combine bitmap graphics inserted in an Excel
sheet as JPG, GIF, or PNG images, with the vector graphics available in Excel
as Shapes. In the following example, a TIF file of the world sea map is
combined with an SVG file (scalable vector graphics), which contains the
contours of continents and islands in a vector format (Fig. 4.10). Although it
contains vector graphic information, this file is essentially textual, can be
opened as such with the Notepad text editor, and contains the coordinates of
the contour points.

    Create the procedures and functions to solve this problem to combine
raster and vector graphics.

Table 4.9 Procedures *Insert_Image()*, *Insert_OpenPolyline()*, *GeoCoord_To_Pixels()*, saved in Module1 in Ex04-10.xlsm

```
Sub Insert_Image(ByVal SheetStr$, ByVal FileName$)
Dim W As Worksheet
  Set W = Application.Worksheets(SheetStr): W.Activate
  W.Range("A1").Select
  W.Pictures.Insert(FileName).Select
End Sub

Sub DrawPolyline(SheetStr$, X() As Double, Y() As Double, N1 As Long, N2 As Long, Rl%, Gl%, Bl%, Rf%, Gf%, Bf%,
Weight%, Closed%)
Dim i&, W As Worksheet
  Set W = Application.Worksheets(SheetStr): W.Activate
  With W.Shapes.BuildFreeform(msoEditingAuto, X(N1), Y(N1))
    For i = N1 + 1 To N2
      .AddNodes msoSegmentLine, msoEditingAuto, X(i), Y(i)
    Next
    If Closed = 1 Then .AddNodes msoSegmentLine, msoEditingAuto, X(N1), Y(N1)
    .ConvertToShape.Select
  End With
  With Selection.ShapeRange.Line
    .Visible = msoTrue
    .ForeColor.RGB = RGB(Rl, Gl, Bl)
    .Weight = Weight
    .Transparency = 0
  End With
  If Closed = 1 Then
    With Selection.ShapeRange.Fill
      .Visible = msoTrue
      .ForeColor.RGB = RGB(Rf, Gf, Bf)
      .Transparency = 0
      .Solid
    End With
  End If
  Range("A1").Select
End Sub
```

Solution

All points in the contours of the continents are stored in 2 arrays $X()$ and $Y()$. Information about the numbers of the start and end points of continents' contours is also stored. The solution is demonstrated in the file Ex04-10.xlsm. Module1 includes a combination of drawing procedures (see Table 4.9), and some of them are already mentioned—for instance, *Delete_AllShapes_Sheet ()*:

- The procedure *DrawPolyline()* has arguments: *SheetStr* is the sheet name, where the line will be displayed, two linear arrays with coordinates $X()$ and $Y()$, *N1* and *N2* are the first and last number of the points, which shape the closed polyline, *Rl, Gl, and Bl* are the RGB components of the contour line of the polyline, and *Rf, Gf, and Bf* are the RGB components of its fill color, *Weight* is the thickness of the polyline.
- The procedure *Insert_Image()* inserts an image (of type.JPG,.BMP,.TIF) into the specified sheet *SheetStr* in position A1.
- The procedure *Read_SVG()* reads the data from SVG file, analyzes them, and draws the contours described in a *left-oriented* coordinate system.

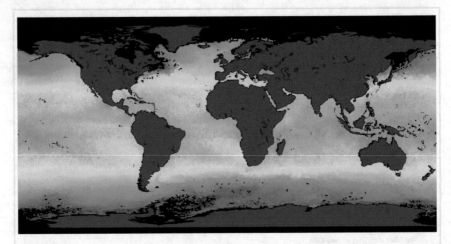

Fig. 4.10 Example for a combination of raster graphic (map) and vector objects, mapped over the image (highways). The image is generated within Ex04-10.xlsm. *Source of the map* (Google Map). The image is generated within Ex04-10.xlsm

The origin of the SVG coordinate system is in the upper left corner. The variable sc is a scale factor to match the image size and the size of contours in the SVG file.

- The procedure *Map_SVG_Main*() calls the above procedures.

Discussion
The insertion of an image in the active sheet is performed with the command *ActiveSheet.Pictures.Insert*(), and it happens in the upper left corner of the current cell (in the case of Fig. 4.10 it is cell A1, which is selected).

The procedure *DrawPolyline*() generates a shape polyline composed of rectilinear sections in a given sheet.

Example 4.1.11 Using other unknown shapes in MS Excel
The list of shapes available in MS Excel is quite large. There are two possible approaches if one wants to use an unknown shape. The first approach is to look for information about it in the documentation of MS Office. There is another more unconventional approach, which sometimes saves a lot of time. For this purpose, the creators of VBA have provided in most products of MS Office the ability to record macros generated in dialog mode. This feature is controlled by two main buttons: [Record Macro] and [Stop Recording], both in Developer Tab.

Table 4.10 Procedure *NewShape()*.

Sub NewShape()
'
' NewShape Macro
'
   ActiveSheet.Shapes.AddShape(msoShapeCube, 125, 95, 150, 210).Select
End Sub

The steps to use this helpful option are as follows:

1. Click the *Developer* tab.
2. In the *Code* group, click on the *Macro* button.
3. In the *Record Macro* dialog box, enter the name of your macro, for instance, *NewShape*.
4. In the '*Store macro in*' option, select '*This Workbook*'.
5. Click *OK*. This will start to record your actions in Excel. The inscription '*Stop recording*' in the *Developer* tab indicates the macro recording is in progress.
6. Select *Insert* Tab, *Shapes*, Main Shapes > *Cube*
7. Draw the cube.
8. Click on the *Stop Recording* button in the *Developer* tab.
9. Open VBA Editor and go to the code of the new macro Sub *NewShape()*.

The automatically recorded macro code in the VBA editor should be something as in Table 4.10.

According to this record, the name of the shape object is *msoShapeCube*, and it needs four arguments—X and Y coordinates in points of the upper left and bottom right corners of the cube frame. This knowledge could help us to create more complex graphic compositions of cubes and parallelepipeds.

4.2   Geometric Calculations with VBA

Most engineering tasks solve problems that essentially occur in three-dimensional space. In many cases, these solutions are visualized with two-dimensional drawings. Most people generally accept that MS Excel is software mostly used for data processing, but this data can also be geometric.

In this sub-section, we will deal with some basic geometric problems in the plane and three-dimensional space.

Example 4.2.1 Distance between two points in a plane or 3D space
Problem definition
A common component of many geometric tasks is calculating the distance between two points in the plane given by their two-dimensional coordinates (x_1, y_1) and (x_2, y_2). Prepare two functions to estimate the distance between two points in a plane and 3D space.
Solution
The Pythagorean theorem can help to solve this problem, according to which the square of the hypotenuse of a right triangle is equal to the sum of the squares of the triangle's legs. For this reason, the required distance L between the two points (Fig. 4.11) is calculated by Eq. (4.1):

$$L = \sqrt{(x_2 - x_1)^2 + (y_2 - y_1)^2} \qquad (4.1)$$

The solution is demonstrated in the file Ex04-11.xlsm. Module1 includes the following functions and a procedure (see Table 4.11):

- The function *Distance_2D()* estimates the distance between two points with coordinates $(\times 1, y1)$ and $(\times 2, y2)$ in the plane.
- The function *Distance_3D()* estimates the distance between two points with coordinates $(\times 1, y1, z1)$ and $(\times 2, y2, z2)$ in the space.
- The sub-routine *Main()* calls above functions.

Discussion
The keyword **ByVal** before each of the six arguments in the declaration of the above functions allows submitting variables or constants of any numerical types—Integer, Long, Single, Double. The result value of the function is of the Double type, which is recommended for most computer graphics tasks to achieve greater accuracy of geometric calculations.

Fig. 4.11 Determining the distance between two points by the Pythagorean theorem

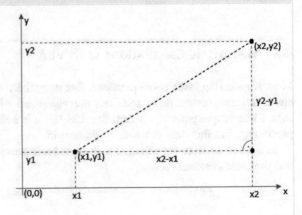

Table 4.11 Functions *Distance_2D()* and *Distance_3D()*, saved in Module1 in Ex04-11.xlsm

Function Distance_2D(ByVal x1#, ByVal y1#, ByVal x2#, ByVal y2#) As Double
'estimate distance between two 2D-points
    Distance_2D = Sqr((x2 - x1) ^ 2 + (y2 - y1) ^ 2)
End Function

Function Distance_3D(ByVal x1#, ByVal y1#, ByVal z1#, _
ByVal x2#, ByVal y2#, ByVal z2#) As Double
'estimate distance between two 3D-points
    Distance_3D = Sqr((x2 - x1) ^ 2 + (y2 - y1) ^ 2 + (z2 - z1) ^ 2)
End Function

Calling the first function for calculating the distance between points with coordinates (2,4) and (6,7) is as follows:
D = Distance_2D(2,4, 6,7)
Calling the second distance calculation function between two points with three-dimensional coordinates (2,4,1) and (6,7,2) looks like this:
D = Distance_3D(2,4,1, 6,7,2)

Example 4.2.2 Perimeter and area of a triangle in a plane and 3D space with Heron's formula
Problem definition
A practical application of both functions shown above in Example 4.2.1 is when calculating the perimeter of a triangle (Fig. 4.12).

Create a function to estimate the perimeter and the area of a triangle, whose vertices are given by their coordinates in a plane or in space.
Solution
Let's create a function to determine the perimeter of a triangle in the plane, which receives as arguments the two-dimensional coordinates (x_1, y_1), (x_2, y_2), and x_3, y_3 and returns a result of type Double for the length of the perimeter. The lengths of each side a, b, and c and triangle's perimeter p are calculated by the following equations—from (4.2) to (4.5):

$$a = \sqrt{(x_2 - x_1)^2 + (y_2 - y_1)^2} \qquad (4.2)$$

$$b = \sqrt{(x_2 - x_3)^2 + (y_2 - y_3)^2} \qquad (4.3)$$

$$c = \sqrt{(x_3 - x_1)^2 + (y_3 - y_1)^2} \qquad (4.4)$$

Fig. 4.12 Drawing of a triangle in the plane to determine its perimeter and area

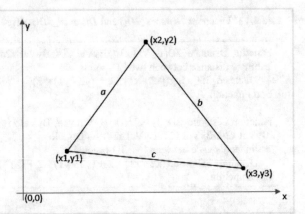

$$p = a + b + c \tag{4.5}$$

The logical next step is to prepare a function so that we can calculate the triangle area. One possible solution is to the Heron's formula (4.6):

$$S = \sqrt{p(p-a)(p-b)(p-c)} \tag{4.6}$$

using the calculated lengths of the sides a, b, and c of the triangle and semi-perimeter $p = (a+b+c)/2$.

Such function is relatively to adapt for calculations in space.

The solution is demonstrated in the file Ex04-12.xlsm. Module1 includes the already mentioned functions *Distance_2D()* and *Distance_3D()* and following new procedures (see Table 4.12):

- The function *Perimeter_2D_triangle()* estimates the perimeter of a triangle, whose coordinates are $(\times 1,y1)$, $(\times 2,y2)$, and $(\times 3,y3)$ in the plane.
- The function *Area_2D_triangle_Heron()* estimates the area of a triangle, whose coordinates are $(\times 1,y1)$, $(\times 2,y2)$, and $(\times 3,y3)$ in the plane. The function uses Heron's formula.
- The function *Area_3D_triangle_Heron()* estimates the area of a 3D triangle, whose coordinates are $(\times 1,y1,z1)$, $(\times 2,y2,z2)$, and $(\times 3,y3,z3)$ in the space. The function uses Heron's formula.
- The sub-routine *Main()* calls above functions.

Discussion

Let's look at examples of how the three functions shown in Table 4.12 can be called and used. Calling the first function to calculate the perimeter of a triangle with coordinates (2,4), (6,7), and (6,4) is as follows:

P = Perimeter_2D_triangle(2,4, 6,7, 6,4)

Table 4.12 Functions *Perimeter_2D_triangle()*, *Area_2D_triangle_Heron()*, *Area_3D_triangle_Heron()*, saved in Module1 in Ex04-12.xlsm

```
Function Perimeter_2D_triangle(ByVal x1#, ByVal y1#, ByVal x2#, ByVal y2#, _
        ByVal x3#, ByVal y3#) As Double
Dim a#, b#, c#
'estimation of edges' lengths
  a = Distance_2D(x1, y1, x2, y2)
  b = Distance_2D(x2, y2, x3, y3)
  c = Distance_2D(x1, y1, x3, y3)
'estimation of perimeter
  Perimeter_2D_triangle = a + b + c
End Function


Function Area_2D_triangle_Heron(ByVal x1#, ByVal y1#, ByVal x2#, ByVal y2#, _
        ByVal x3#, ByVal y3#) As Double
Dim a#, b#, c#, P#
'estimation of edges' lengths
  a = Distance_2D (x1, y1, x2, y2)
  b = Distance_2D (x2, y2, x3, y3)
  c = Distance_2D (x1, y1, x3, y3)
'estimation of semi-perimeter
  P = (a + b + c) / 2
  Area_2D_triangle_Heron = Sqr(P * (P - a) * ( P - b) * (P - c))
End Function


Function Area_3D_triangle_Heron(ByVal x1#, ByVal y1#, ByVal z1#, _
        ByVal x2#, ByVal y2#, ByVal z2#, _
        ByVal x3#, ByVal y3#, ByVal z3#) As Double
Dim a#, b#, c#, P#
'estimation of edges' lengths
  a = Distance_3D(x1, y1, z1, x2, y2, z2)
  b = Distance_3D(x2, y2, z2, x3, y3, z3)
  c = Distance_3D(x1, y1, z1, x3, y3, z3)
'estimation of semi-perimeter
  P = (a + b + c) / 2
  Area_3D_triangle_Heron = Sqr(P * (P - a) * (P - b) * (P - c))
End Function
```

Calling the second function to calculate the area of a triangle with coordinates (2,4), (6,7), and (6,4) is similar:

S = Area_2D_triangle_Heron(2,4, 6,7, 6,4)

When calling the 3D version of this function, the function needs to get nine coordinates, three for each point, for example:

S = Area_3D_triangle_Heron(2,4,1, 6,7,2, 6,4,5)

Example 4.2.3 Area of a triangle in a plane and 3D space with vector product
Problem definition
The problem with the last two functions in Example 4.2.2 is that they use relatively slow operations. These are the three calls to the *Sqr()* function for the square root. Some engineering tasks could call the area function thousands of times. A much simpler and shorter solution exists, which, however, requires a little more mathematical knowledge.

In addition to the Heron's formula, we can calculate the face of a triangle with a *vector product*. Like most *ingenious things*, the equation for this is longer to explain than to write.

Let's explain how to calculate the *vector product of two vectors*, each of which is formed on one of the triangle sides whose area we are looking for (Fig. 4.13). The vector product of two such vectors $\vec{a}$ and $\vec{b}$ is a third vector $\vec{a} \times \vec{b}$, which is perpendicular to the plane of the two vectors and of length corresponding to the face of the parallelogram formed by the two vectors (Fig. 4.13).

$$\vec{a} \times \vec{b} = \begin{vmatrix} \vec{i} & \vec{j} & \vec{k} \\ a_1 & a_2 & a_3 \\ b_1 & b_2 & b_3 \end{vmatrix} = \begin{vmatrix} \vec{i} & \vec{j} & \vec{k} \\ a_1 & a_2 & 0 \\ b_1 & b_2 & 0 \end{vmatrix} \tag{4.7}$$

$$\vec{a} \times \vec{b} = \vec{k} \begin{vmatrix} a_1 & a_2 \\ b_1 & b_2 \end{vmatrix} = \vec{k} \begin{vmatrix} x_2 - x_1 & y_2 - y_1 \\ x_3 - x_1 & y_3 - y_1 \end{vmatrix} \tag{4.8}$$

If $\vec{a}$ is the vector from point 1 $(x_1, y_1, 0)$ to point 2 $(x_2, y_2, 0)$, and $\vec{b}$ is the vector from point 1 $(x_1, y_1, 0)$ to point 3 $(x_3, y_3, 0)$, then the vector $\vec{a}$ has coordinates $(x_2-x_1, y_2-y_1, 0)$, and the vector $\vec{b}$ has coordinates $(x_3-x_1, y_3-y_1, 0)$.

Thus, the vector product is calculated as a coefficient of the vertical unit vector $\vec{k}$ (Eq. 4.7 and 4.8) and is equal to $a_1*b_2 - b_1*a_2$, i.e. $(x_2-x_1) * (y_3-y_1) - (x_3-x_1) * (y_2-y_1)$, and the area of the triangle is half of the absolute value of the result.

Fig. 4.13 Definition of the vector product, image source: (**Wikipedia— Vector product**)

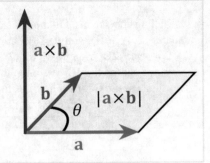

Table 4.13 Function *Area_2D_triangle_vector_product()*, saved in Module1 in Ex04-13.xlsm

Function Area_2D_triangle_vector_product(ByVal x1#, ByVal y1#, _
                              ByVal x2#, ByVal y2#, _
                              ByVal x3#, ByVal y3#) As Double
    Area_2D_triangle_vector_product = Abs((x2 - x1) * (y3 - y1) _
      - (x3 - x1) * (y2 - y1)) / 2
End Function

Create a function for calculating the area of a triangle using a vector product.

Solution

The solution is demonstrated in the file Ex04-13.xlsm. Module1 includes the following procedures (see Table 4.13):

- The function *Area_2D_triangle_vectorproduct()* estimates the area of a 2D triangle, whose coordinates are $(\times 1, y1)$, $(\times 2, y2)$, and $(\times 3, y3)$ in the plane. The function uses a vector product and is very short.
- The sub-routine *Main()* calls above function.

Discussion

An example of calling the *obvious much more compact and faster function* for the estimation of triangle area by a vector product formed by *three consecutive set points*:

MsgBox("Area = " & Area_2D_triangle_vector_product(2,4, 6,7, 6,4))

This solution is especially good if it is to be applied as part of more complex problems and in the repeated calculation of faces of many triangles.

Example 4.2.4 Area of convex and concave irregular polygons

Problem definition

We can apply the functions of calculating the face of a triangle to calculate the face of a non-self-intersecting polygon given by the coordinates of its **N** vertices. For a convex polygon, the problem can be easily solved using the triangle area function according to the Heron's formula. The area of the polygon is estimated as a sum of (N-2) areas of triangles, formed when the first vertex of the polygon is connected to its other vertices (from the third to the penultimate).

The approach that uses Heron's formula is not suitable for a concave polygon. In this case, the most appropriate approach is to use the function to

Table 4.14 Functions *Area_IrregularPolygon_Heron*() and *Area_IrregularPolygon_Vector*(), saved in Module1 in Ex04-14.xlsm

```
Function Area_IrregularPolygon_Heron(x() As Double, y() As Double, N%) As Double
Dim S As Double, i As Integer
'estimation of polygon's area, using Heron's formula
 S = 0
 For i = 1 To N - 2
   S = S + Area_2D_triangle_Heron(x(1), y(1), x(i + 1), y(i + 1), _
     x(i + 2), y(i + 2))
 Next
 Area_IrregularPolygon_Heron = S
End Function

Function Area_IrregularPolygon_Vector(x() As Double, y() As Double, N%) As Double
Dim S As Double, i As Integer
'estimation of polygon's area, using vector product
 S = 0
 For i = 1 To N - 2
   S = S + Area_2D_triangle_vector_product (x(1), y(1), x(i + 1), y(i + 1), _
     x(i + 2), y(i + 2))
 Next
 Area_IrregularPolygon_Vector = Abs(S)
End Function
```

calculate the area of a triangle with a vector product. The function uses the *Area_2D_triangle_vector_product*() function defined above. For a convex polygon with a high number of vertices, this function will work much faster.

Create a function to estimate the area of irregular convex polygons, using the Heron's formula, and another function for estimating the area of irregular concave polygons, using vector product.

Solution

The solution is demonstrated in the file Ex04-14.xlsm. Module1 includes the already mentioned functions *Distance_2D*(), *Area_2D_triangle_Heron*() and *Area_2D_triangle_vectorproduct*(), and following procedures (see Table 4.14):

- The function *Area_IrregularPolygon_Heron*() estimates the area of a convex 2D irregular polygon, whose vertices are passed as arrays x() and y(), and N is the number of vertices. The function is not suitable for concave polygons.
- The function *Area_IrregularPolygon_Vector*() estimates the area of a convex of concave2D irregular polygon, whose vertices are passed as arrays x() and y(), and N is the number of vertices.
- The sub-routine *Main*() calls above function.

Discussion
The coordinates of the points are passed to both functions as two one-dimensional arrays of type Double in combination with the argument N, which is the number of polygon vertices.

4.3 Geometric Transformations of Objects in the Plane

When modeling more complex geometric compositions, the following transformations are useful—translation, rotation, and resizing of objects determined by the coordinates of their characteristic points in the plane. In the mentioned geometric transformations, only the geometric information changes, and the topological information, which shows the way of connecting the points between them, remains the same and does not change.

Example 4.3.1 Translation of a geometric object in the plane
Problem definition
Let a plane object (e.g. a closed or non-closed contour composed of linear segments) is defined by the coordinates of its characteristic points, written in two arrays $Xc()$ and $Yc()$, each with N elements of type Double. Two numbers set the translation vector—its projections dx and dy on the X and Y axes (Fig. 4.14).

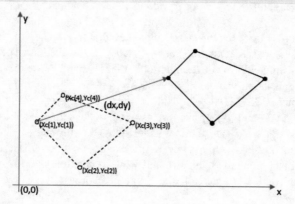

Fig. 4.14 Illustration for translation of a geometric object in the plane. The object before the translation is depicted with a dotted line, and after it—with a solid line. The image is generated within Ex04-15.xlsm

Table 4.15 Procedure *Move_2DObject*(), saved in Module1 in Ex04-15.xlsm

```
Sub Move_2DObject(ByRef Xc() As Double, ByRef Yc() As Double, _
    ByVal N As Integer, ByVal dx As Double, ByVal dy As Double)
Dim i%
  For i = 1 To N
    Xc(i) = Xc(i) + dx
    Yc(i) = Yc(i) + dy
  Next
End Sub
```

Create a procedure that recalculates the coordinates of the object's points after translation and another that draws the geometric object before and after translation.

Solution

The location of the individual points changes according to Eqs. (4.9) and (4.10), where xc and yc are coordinates of the points before translation, and xc' and yc'—after it.

$$xc' = xc + dx \qquad (4.9)$$

$$yc' = yc + dy \qquad (4.10)$$

The solution is demonstrated in the file Ex04-15.xlsm. The coordinates of the object's points are placed in sheet Data and vector coordinates (dx, dy). Module1 includes the already mentioned sub-routines *DrawLine*(), *DrawArrow*(), *Delete_AllShapes_Sheet*(), *DrawRectangle*(), *DrawCircle*(), *InsertText*(), and the following procedures (see Table 4.15):

- Procedure *Move_2DObject*() recalculates the coordinates of the object's vertices, stored in X() and Y() arrays, (×00, y00) is the left upper corner of the drawing field, N is the number of vertices, dx and dy are the coordinates of the translation vector.
- Procedure *Draw_2DObject*() draws the object, using object's vertices, stored in X() and Y() arrays, (×00, y00) is the left upper corner of the drawing field, N is the number of vertices, the meaning of the variables *Style* and *Dashed* is the same as in Example 4.1.9, the variable *Ctype* draws black points for value 0, and white points for 1.
- The sub-routine *Main*() calls the above procedures.

Discussion

The procedure is relatively simple. We can call it as follows:

Call Move_2DObject (Xc(), Yc(), N, A, B)

to move the object at a distance of *A* units on the X-axis and *B* units on the Y-axis, where *A* and *B* could be positive or negative numbers. The *Move_2DObject*() procedure receives the two coordinate arrays as arguments at an address with the keyword *ByRef* to return their recalculated values to the calling procedure. The parameter N for the number of points defining the geometric object is also passed; it determines how many pairs of coordinates should be processed in the procedure.

Example 4.3.2 Scaling (homothety) of a geometric object in the plane
Problem definition
The next task is to increase or decrease an object with a given scale factor *ScaleXY* relative to a point with coordinates (*×0, y0*), which is the center of homothety (Fig. 4.15).
Create procedures to perform the following tasks:

1. Recalculate the coordinates of a given point when scaling with a given scale factor *ScaleXY* relative to the center of homothety (*×0, y0*);
2. Recalculate the coordinates of an object's characteristic points, scaled with a scale factor *ScaleXY* in regard to the center of homothety (*×0, y0*).

Solution
The solution is demonstrated in the file Ex04-16.xlsm. The coordinates of the object's points are placed in sheet Data, the coordinates of the center of homothety (*×0,y0*), and scale factor Scale_XY. Module1 includes the already

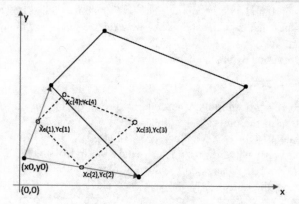

Fig. 4.15 Scaling a geometric object in the plane. The center of scaling is point (x0, y0), and the scaling factor is 2. The object before scaling is depicted with a dotted line, and after it—with a solid line. The image is generated within Ex04-16.xlsm

Table 4.16 Procedures *Scale_2DPoint*() and *Scale_2DObject*(), saved in Module1 in Ex04-16. xlsm

```
Sub Scale_2DPoint(ByRef xc As Double, ByRef yc As Double, _
     ByVal x0 As Double, ByVal y0 As Double, ByVal ScaleXY As Double)
Dim x#, y#
  x = xc - x0: y = yc - y0
  xc = x0 + x * ScaleXY
  yc = y0 + y * ScaleXY
End Sub

Sub Scale_2DObject(ByRef Xc() As Double, ByRef Yc() As Double, ByVal N As Integer, _
     ByVal x0 As Double, ByVal y0 As Double, ByVal ScaleXY As Double)
Dim i%
  For i = 1 To N
    Call Scale_2DPoint(Xc(i), Yc(i), x0, y0, ScaleXY)
  Next
End Sub
```

mentioned sub-routines *DrawLine*(), *DrawArrow*(), *Delete_AllShapes_Sheet* (), *DrawRectangle*(), *DrawCircle*(), *InsertText*(), *Draw_2DObject*(), and the following new procedures (see Table 4.16):

- The procedure *Scale_2DPoint*() gets as arguments the coordinates (*xc, yc*) that have to be recalculated, then coordinates (×*0, y0*) of the homothety's center, to which the operation is performed, and finally, the factor *ScaleXY*.
- The procedure *Scale_2DObject*() does the same as *Scale_2DPoint*(), but for all points of the object. The coordinates of the points are given as two one-dimensional arrays, accompanied by the argument *N*, which is the number of points. If *ScaleXY* is greater than 1, the object increases, and if it is between 0 and 1, it decreases. The scale factor can also be less than 0. Investigate what happens at *ScaleXY* between 0 and -1, as well as at values less than -1.
- The sub-routine *Main*() calls the above procedures.

Discussion
Calling the last procedure can be done in the following way to double the magnification of the object:
  Call Scale_2DObject(Xc(), Yc(), N, × 0, y0, 2)
  It is relatively easy to rework the procedure to use scale factors different on axes X and Y—*ScaleX* and *ScaleY*. Such a solution allows you to change the proportions of the object on both axes when needed.

Example 4.3.3 Rotation of a geometric object in the plane

Problem definition

Another frequently performed task on objects in plane modeling is rotation, i.e. rotation at an angle of an object given by the coordinates of its characteristic points. If the rotation angle *Alpha* is in degrees, it is necessary to convert it into radians, using the equation $\pi = 180°$. Let the rotation be performed relative to a point with coordinates $(x0, y0)$, and the positive direction for measuring the angle is counter clockwise, as in the right-hand coordinate system. If we want to rotate the object clockwise, we should specify a negative rotation angle (Fig. 4.16).

In order to calculate the new coordinates xc' and yc' at each rotated point, it is necessary to calculate the projections x and y before rotating the directional segment from point $(x0, y0)$ to point (xc, yc) on the X and Y axes. Their lengths are calculated as $x = xc - x0$ and $y = yc - y0$. In this case, dx and dy, which are the projections of the same segment after the rotation of the angle α, are calculated using formulas (4.11) and (4.12), and the new coordinates of the point (xc, yc) are calculated using Eqs. (4.13) and (4.14):

$$dx = x \cos\alpha - y \sin\alpha \tag{4.11}$$

$$dy = x \sin\alpha + y \cos\alpha \tag{4.12}$$

$$xc' = x0 + dx \tag{4.13}$$

$$yc' = y0 + dy \tag{4.14}$$

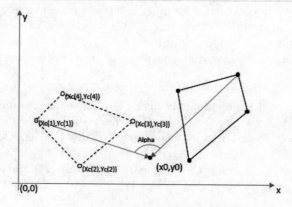

Fig. 4.16 Rotation of a geometric object in the plane. In this example, the rotation angle α is negative ($-120°$), and therefore, the rotation is clockwise. The object before the operation is depicted with a dotted line, and after it—with a solid line. The image is generated within Ex04-17.xlsm

Table 4.17 Function *Degrees_To_Radians*() and procedures *Rotate_2DPoint*() and *Rotate_2-DObject*(), saved in Module1 in Ex04-17.xlsm

```
Function Degrees_To_Radians(ByVal Alpha As Double) As Double
   Degrees_To_Radians = Alpha / 180 * Application.Pi()
End Function

Sub Rotate_2DPoint(ByRef xc As Double, ByRef yc As Double, _
      ByVal x0 As Double, ByVal y0 As Double, ByVal Alpha As Double)
Dim x#, y#, Alpha_Rad#
   Alpha_Rad = Degrees_To_Radians(Alpha)
   x = xc - x0: y = yc - y0
   xc = x0 + x * Cos(Alpha_Rad) - y * Sin(Alpha_Rad)
   yc = y0 + x * Sin(Alpha_Rad) + y * Cos(Alpha_Rad)
End Sub

Sub Rotate_2DObject(ByRef Xc() As Double, ByRef Yc() As Double, ByVal N As Integer, _
      ByVal x0 As Double, ByVal y0 As Double, ByVal Alpha As Double)
Dim i%
   For i = 1 To N
      Call Rotate_2DPoint(Xc(i), Yc(i), x0, y0, Alpha)
   Next
End Sub
```

Create procedures to perform the following tasks:

1. Convert the angle *Alpha*, given in degrees, to radians.
2. Calculate the coordinates of a point with coordinates (*xc, yc*) rotated at an angle *Alpha* relative to the point (×0, y0).
3. Calculate the new coordinates of the points of a plane object rotated at an angle *Alpha* relative to the center of rotation (×0, y0). The coordinates of the points should be submitted to the procedure as two one-dimensional arrays, in combination with the number of points defining the object.

Solution

The solution is demonstrated in the file Ex04-17.xlsm. The coordinates of the object's points are placed in sheet Data, the coordinates of the center of rotation (×0,y0), and rotation angle Alpha. Module1 includes the already mentioned sub-routines *DrawLine*(), *DrawArrow*(), *Delete_AllShapes_Sheet* (), *DrawRectangle*(), *DrawCircle*(), *InsertText*(), *Draw_2DObject*(), and the following new procedures (see Table 4.17):

- The function *Degrees_To_Radians*() is for converting an angle from degrees to radians. In it, the value of π is obtained by.*Pi*() method of the application object, which returns this mathematical constant with an accuracy of 15 digits.

- The procedure *Rotate_2DPoint()* performs the necessary calculations for the rotation of a point with coordinates (*xc, yc*) relative to a point (×*0, y0*) at an angle *Alpha* (set in degrees). Because the trigonometric functions *Cos* () and *Sin()* require arguments in radians, we have to convert the angle *Alpha* to a new variable named *Alpha_Rad* in radians using the previous function *Degrees_To_Radians()*.
- The procedure *Rotate_2DObject()* calculates the new coordinates of all characteristic points of a plane object, written in the two arrays *Xc()* and *Yc* (), after rotation relative to the point (×*0, y0*) at an angle *Alpha* (set in degrees). The procedure repeatedly calls the above function *Rotate_2DPoint()* to calculate the new location after the rotation of each characteristic point of the object.
- The sub-routine *Main()* calls the above procedures.

Discussion
Finally, there is an example of calling the last procedure *Rotate_2DObject()* rotate the object with *N* coordinates *Xc()* and *Yc()* relative to the center of rotation with coordinates (5,7) at an angle of 45° in a counterclockwise direction.

    Call Rotate_2DObject(Xc(), Yc(), N, 5, 7, 45)

Example 4.3.4 Translation, scaling, and rotation of a geometric object in the plane
Problem definition
It is quite possible that a set of transformations—translation, scaling, and rotation—will need to be applied to an object (Fig. 4.17).

    Use the procedures defined above to perform for an existing planar object:

1. Translation with vector (*TranslationX, TranslationY*);
2. Scaling relative to a specified point (*Xcenter, Ycenter*) with scale factor *ScaleXY*;
3. Rotate Angle *RotationAngle_Deg* relative to the same point (*Xcenter, Ycenter*).

Solution
The solution is demonstrated in the file Ex04-18.xlsm. The coordinates of the object's points are placed in sheet Data, and also the coordinates of the translation vector, the coordinates of the scaling center and the scale factor, and the coordinates of rotation center (×*0,y0*) and rotation angle *Alpha*.

    Module1 includes the already mentioned sub-routines *DrawLine()*, *DrawArrow()*, *Delete_AllShapes_Sheet()*, *DrawRectangle()*, *DrawCircle()*,

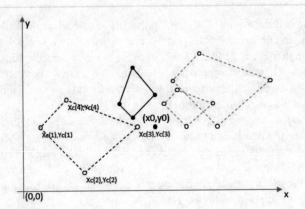

Fig. 4.17 Combination of translation, scaling, and rotation of a geometric object in the plane. The scaling center is (×0, y0), the rotation angle is 45°, and the scaling factor is 0.5. The object before the changes has a black dotted line, and after them—with a solid black line. The intermediate states of the object are depicted in gray. The image is generated within Ex04-18. xlsm

InsertText(), *Draw_2DObject*(), *Move_2DPoint*(), *Move_2DObject*(), *Scale_2DPoint*(), *Scale_2DObject*(), *Degrees_To_Radians*(), *Rotate_2DPoint*(), *Rotate_2DObject*(), and the procedure *Main*() that calls the above procedures.

The three procedures defined above should be called one after the other to perform the three tasks.

Call Move_2DObject(Xc(), Yc(), N, TranslationX, TranslationY)

Call Scale_2DObject(Xc(), Yc(), N, Xcenter, Ycenter, ScaleXY)

Call Rotate_2DObject(Xc(), Yc(), N, Xcenter, Ycenter, RotationAngle_Deg)

The procedures and functions listed in Sect. 4.3 can allow us to perform any operations that change the size, location, and orientation of objects in the plane.

Example 4.3.5 Transformation from absolute to user coordinate system in the plane

Problem definition

Many engineering tasks need to visualize two-dimensional objects by switching from one absolute two-dimensional coordinate system (in which the coordinates of all objects are set) to another user coordinate system. One possible example is to enlarge a part of a drawing on the screen to occupy the whole drawing field (zooming)—in a dotted line in Fig. 4.18.

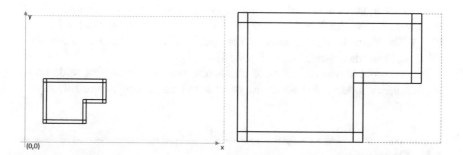

Fig. 4.18 Switching from absolute to relative coordinate system and scaling the image: **a** before zooming; **b** after zooming. The drawing field is marked with a dashed gray line. The image is generated within Ex04-19.xlsm

Use the procedures defined above to find the relative coordinates of an object relative to a new origin of a custom coordinate system whose coordinates in the absolute coordinate system are (*Xnew, Ynew*). For this purpose, do the following actions:

1. Translation of the coordinate system with a vector (−*Xnew*, −*Ynew*);
2. Scale the objects so that they occupy the entire drawing field.

Solution
The solution is demonstrated in the file Ex04-19.xlsm. The coordinates of the object's points are placed in sheet Data.

In order to perform the two tasks, the two procedures defined above for working with objects in the plane—for translation and for scaling—must be called one after the other as follows:

Call Move_2DObject(Xc(), Yc(), N, -Xnew, -Ynew)
Call Scale_2DObject(Xc(), Yc(), N, 0,0, ScaleXY)

Discussion
When calling the first procedure, the values of *Xnew* and *Ynew* must correspond to the coordinates of the lower-left corner of the object, i.e. of the minimum values *Xmin* and *Ymin* of the arrays *Xc*() and *Yc*(). When calling the procedure, *Move_2DObject*(), keep in mind that the displacement vector must include sign minus before *Xnew* and *Ynew*.

When calling the second procedure, it is important to take into account that the scaling center after the first procedure is already at the origin of the new coordinate system, i.e. (0,0), where the lower-left corner of the object is located.

An interesting question is how to determine the scale factor *ScaleXY* so that the object fills the drawing field. Let the dimensions of this field are *SizeX* on the X-axis and *SizeY* on the Y-axis. Let the size of the object in the horizontal direction (on the X-axis) is (*MaxX* - *MinX*), and on the Y-axis is (*MaxY* - *MinY*). We determine the scale factor on the X-axis *ScaleX* by the equation $ScaleX = SizeX / (MaxX - MinX)$ and the factor on the Y-axis *ScaleY* by the equation $ScaleY = SizeY / (MaxY - MinY)$. The value of *ScaleXY* must be the smaller of the two factors for the whole object to fit in the drawing field (Fig. 4.18).

It is essential to add that all these transformations are not suitable to do on the original metric information about the objects, but on its copy, intended only for visualization.

4.4 Conclusion

The purpose of this chapter was to acquaint the reader with less popular features of MS Excel—to generate a variety of graphics, beyond the traditional charts. We have shown how two types of raster graphics can be generated, using different approaches: (i) the 'pixels' are generated by coloring cells, and (ii) the 'pixels' are objects from the Shapes collection, available in all products of the MS Office package. Raster graphics of this type have numerous applications. Here are shown maps depicted in Excel, generated from NASA datasets, as well as variations in the view factor at different points from one surface to another. In addition to raster graphics capabilities, a vector graphics application is demonstrated, again using objects from the Shapes collection. In this way, vector legends can be added to the raster images previously demonstrated, as well as vector objects with complex contours read from SVG files can be drawn on them. The second part of this chapter demonstrated how to perform various transformations (displacement, scaling, rotation, and combinations thereof) on two-dimensional objects, as well as generating images visualizing these transformations. All of these examples can greatly expand the capabilities of programmers using VBA in MS Office applications.

References

Google Map, available at: https://maps.google.com/

T. Muneer, S. Ivanova, Y. Kotak, M. Gul, Finite-element view-factor computations for radiant energy exchanges. J. Renew Sustainable Energy **7**, 033108 (2015)

NEO—NASA Earth Observation website, Albedo map, available at: https://neo.sci.gsfc.nasa.gov/view.php?datasetId=MCD43C3_M_BSA&year=2017

NEO—NASA Earth Observation website, available at: https://neo.sci.gsfc.nasa.gov/

NEO—NASA Earth Observation website, Average land surface temperature [day], available at: https://neo.sci.gsfc.nasa.gov/view.php?datasetId=MOD_LSTD_CLIM_M

NEO—NASA Earth Observation website, Average sea surface temperature, available at: https://neo.sci.gsfc.nasa.gov/view.php?datasetId=MYD28M&year=2003

NEO—NASA Earth Observation website, Daily average Land surface temperature [day], available at: https://neo.sci.gsfc.nasa.gov/view.php?datasetId=MOD_LSTD_M&year=2021

NEO—NASA Earth Observation website, Topography, available at: https://neo.sci.gsfc.nasa.gov/view.php?datasetId=SRTM_RAMP2_TOPO

Wikipedia—Vector product, available at: https://en.wikipedia.org/wiki/Cross_product

Dynamics

5

The automobile is truly a success story of our times. People are addicted to their vehicles as they travel with speed, comfort, safety, economy, and in an environment that provides pleasure.

Worldwide, there are now well over a billion cars and that vehicle population is increasing every single day.

The local and global pollution produced by the Internal Combustion Engine (ICE) vehicles is its undoing though and major economies around the world have already started to set dates for phasing out of ICE and its replacement by the Battery Electric Vehicle (BEV).

Figures 5.1 and 5.2 respectively compare the overall efficiencies of various engine technologies and various fuels.

5.1 Comparison of ICE and BEV

A study carried out in the United States showed that overall, the ICEV uses 21.5% of total fuel used to propel the vehicle as illustrated in Fig. 5.3. Energy lost due to exhaust, engine cooling, and transmission losses equates to 78.5% of the total fuel energy.

Figure 5.4 illustrates the breakdown of a comparable passenger EV's energy consumption during winter conditions when heating of cabin space is required. The EV eliminates exhaust losses and has fewer mechanical losses due to fewer moving parts in the vehicle. Around 86% of the battery's energy is contributed to traction energy. An additional factor unique to the EV is the regenerative braking system which typically extends the battery capacity by 23%.

Supplementary Information The online version contains supplementary material available at (https://doi.org/10.1007/978-3-030-94085-0_5).

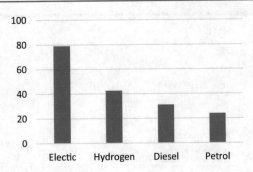

Fig. 5.1 Well-to-wheel efficiencies: comparison of road transport modes

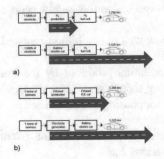

Fig. 5.2 Comparison of mileage delivered by **a** hydrogen and electric vehicle, and **b** ethanol and electricity generated from biomass [Figure redrawn using information presented in Randall (2009)]

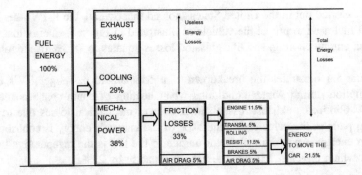

Fig. 5.3 ICE energy losses (redrawn using data presented by Holmerg et al. 2012)

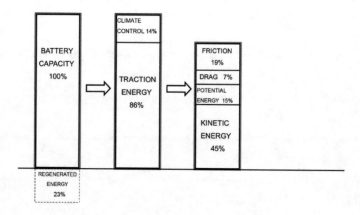

Fig. 5.4 BEV energy losses

5.2 Regenerative Braking Offered by BEV

The facility of regenerative braking goes in favor of Battery Electric Vehicles. In terms of computing energy recovered, one has to know, in addition to the vehicle mass, the initial and final speed. We know from our experience that within the city's environs there are many instances of acceleration and deceleration periods and that has a detrimental effect on fuel efficiency. However, not all braking energy can be recovered. To cite an example, an electric vehicle such as Renault Zoe will not recover kinetic energy when its speed is 15kmph or lower due to the way it has been programmed.

One popular and economic BEV is the Renault Zoe which with one occupant has a total mass of approximately 1600 kg. That vehicle traveling at a speed of 120 km/h (33.33 m/s) has a value of kinetic energy equal to 0.25 kWh. That energy, if recaptured, will enable the vehicle to travel 1.25 km within an urban setting.

The efficiency of regenerative braking-related energy capture depends on the speed though, and Fig. 5.5 demonstrates that relationship.

Figure 5.5 illustrates that when the motor or generator is subjected to a load factor above 0.2. The above data which shows that the efficiency could be as high as 97% is used in the present VEDEC code.

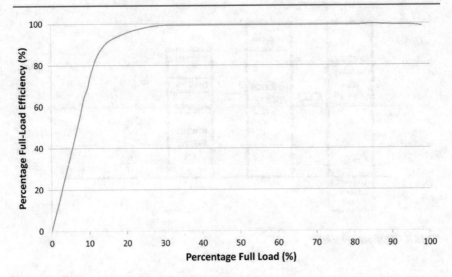

Fig. 5.5 Three-phase induction motor/generator efficiency profile

5.3 Factors That Affect Energy Consumption of BEV

A simulation program will be presented later in this chapter to calculate the energy consumption of the vehicle. The components of the program are:

- Energy used to propel the vehicle. This can be in regard to energy to maintain the speed of the vehicle and also the energy the vehicle consumes to accelerate. To propel the vehicle from stand still is lot more energy intensive than energy required to maintain the vehicle at constant speed. However, as will be discussed later in the chapter, higher constant speeds will have higher energy consumption.
- The gradient the vehicle has to overcome. This will incorporate a change in potential energy.
- Weight of the vehicle including the number of passengers on board.
- Friction between tires and road surface.
- The drag effect. This factor will vary from vehicle to vehicle depending on the vehicle's design that determines the coefficient of drag.
- The impact of the regenerative braking system.

5.4   Driving Cycles

A number of urban test runs were organized by the authors in a BEV with a 24kWh lithium-ion battery capacity. No heating or cooling was utilized during the test run so all energy was used for traction (energy used to propel the vehicle). The route varied in altitude and road surfaces to give real-life data as shown in Fig. 5.6. Real-life data was preferred to be obtained as usually manufacturers will test in laboratories with ideal conditions which can at times be difficult to simulate in real-life conditions and make fair comparisons. The driver was asked to drive in a normal fashion and not be cautious of the energy usage. Table 5.1 represents the data collected from three test runs. The energy consumed by the battery is displayed in miles. This is the amount of energy, the vehicles algorithm computed, that the driver's driving style consumed from the battery range for that trip.

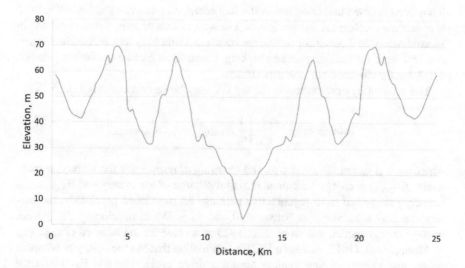

Fig. 5.6 Urban drive cycle topography for city of Edinburgh

Table 5.1 Urban drive cycles

Trial	Duration, mins	Number of times vehicle stopped	Distance Traveled, miles	Energy used by battery, kWh	Energy consumed by battery, miles (deducted from the display range)
1	65	32	5.5	3.98	15
2	62	31	15.5	3.29	3
3	59	31	15.5	3.85	16

Driving styles in urban areas are highly influenced by traffic management infrastructure in place for a particular town or city. As shown in Table 5.1, the driver experienced a very disruptive drive as the vehicle was at a complete stop a significant amount of times. On average, every two minutes the vehicle came to a complete stop at traffic lights. To move the vehicle from stand still consumes a considerable amount of energy. The experiment to demonstrate urban driving showed little difference between experienced and inexperienced EV drivers; although the experienced driver experienced more stops, the results remained very similar to those obtained from the driver not conscious of energy consumption.

5.5 The Vehicle Dynamic and Energy Consumption (VEDEC) Simulation Equations

It has been claimed that one-third of the fuel energy in a conventional ICEV is used to overcome mechanical friction losses. The split of loss of energy due to friction is approximately 35%, each for rolling resistance and tribology loss within the engine, and 15% each for transmission and braking. Presently, a friction coefficient value of 0.013 for the tire-road surface was chosen.

The total energy consumption for the EV may be represented via Eq. (5.1),

$$\text{Total Energy} = \frac{1}{\eta_{\text{bc}}} \left[\frac{1}{\eta_{\text{d}}} E_{\text{Traction}} - \eta_{\text{c}} E_{\text{Regeneration}} \right] \tag{5.1}$$

where η_{d} and η_{c} are the discharge and charging efficiency for the battery, respectively. E_{Traction} is energy consumed during discharge of the battery and $E_{\text{Regeneration}}$ is energy recovered from regenerative braking. As mentioned previously, the battery charging efficiency lies between 90 and 95%. When developing the present traction energy model, this work used 0.925 as a value for the latter efficiency, η_{bc}.

Muneer et al. (2015) present a VEDEC simulation that has the ability to compute power and energy of any vehicle during a drive cycle. The file Ex05-01.xlsm contains the macros of the VEDEC software.

The software can be adopted to the ICEV as well as the BEV. The model may be used as a valuable tool to demonstrate how much energy is saved when moving toward a BEV dominant automobile market. The one advantage that the BEV has over the older ICEV is its ability to capture energy that would be conventionally lost due to braking as aforementioned in this chapter. The software computes energy savings that is recaptured from regenerative braking system when compared directly with the energy requirements of the same vehicle without such system. Various modes of driving styles such as cruise, acceleration, and change of gradient are all considered in the software.

(a) Tire friction component

Friction is a force that resists motion. Friction has been a topic that has attracted a lot of work in the research community. Research is continuously looking at how to reduce tire friction to an optimum yet safe value so that with adequate contact to the ground friction provides motion to propel the vehicle and also gradual deceleration when no traction energy is applied.

Frictional loss may be calculated from Eq. (5.2), where μ is the friction coefficient of tire on the given road surface, m is the mass of the vehicle, and g is the acceleration due to gravity,

$$F_{\text{friction}} = \mu W = \mu mg \qquad (5.2)$$

(b) Potential force component

The energy that an object has due to its position with respect to a datum is incorporated into the simulation. The vehicle's position with respect to a given datum will have a positive or negative gradient angle depending on whether the vehicle is ascending or descending. This force is also dependent on the vehicle's weight and is computed from Eq. 5.3,

$$F_{\text{potential}} = W \sin\theta = mg \sin\theta \qquad (5.3)$$

where θ is the angle relating to the gradient of the road on which the vehicle is traveling.

(c) Aerodynamic drag component

Air drag is another force that automobile designers are trying to combat as effectively as possible. The shape of the vehicle has changed dramatically from the very first steam-powered locomotive to the automobiles on the road today. Altering the shape of the vehicle's shell can increase or reduce the air drag coefficient dramatically. The air drag component considered in the present VEDEC simulation is calculated via Eq. 5.4,

$$F_{\text{drag}} = \frac{1}{2}C_d A \rho v^2 = \frac{1}{2}C_d A \rho \frac{v_f^2 + v_i^2}{2} = \frac{1}{4}C_d A \rho (v_f^2 + v_i^2), \qquad (5.4)$$

where C_d is the drag coefficient of a specific vehicle, A is the frontal cross-sectional area of the vehicle, ρ is the density of air (standard value = 1.225 kg/m^3), v_f is final velocity, and v_i is the initial velocity.

(d) Kinetic energy component

While the vehicle is moving, it builds up kinetic energy. This is energy a vehicle
possesses by virtue of its motion. The change in this energy can be positive or
negative depending on whether the vehicle is accelerating or decelerating. The two
variables that are considered in kinetic energy are the mass and the velocity of the
vehicle (v). Thus, from Eq. 5.5,

$$F_{\text{kinetci}} = \frac{1}{2}mv^2 = \frac{1}{2}m(V_f^2 - V_i^2) \tag{5.5}$$

(e) The complete equation

The Excel-VBA code within the file Ex04-01.xlsm is the VEDEC simulation
software that has been extensively tested using urban, rural, and motorway driving
cycles undertaken by two vehicles: a Renault Zoe and a Nissan Leaf.

Equation 5.6 is used in this software to calculate power and energy requirements
of a vehicle. All of the relevant forces, friction, potential energy, aerodynamic drag
as well as the change in kinetic energy are considered respectively on the right-hand
side of Eq. 5.6 where Δd represents the distance traveled by the vehicle in a given
time interval.

$$E = \left[\mu mg \cos\theta + mg \sin\theta + \frac{1}{4}C_d A\rho(v_f^2 + v_i^2)\right]\Delta d + \frac{1}{2}m(v_f^2 - v_i^2) \tag{5.6}$$

Many of the variables (such as μ, m, C_d, and A) are obtained from manufacturers' specifications for specific vehicle, and others are known constants (such as
g and ρ) or can be obtained from measurements and/or surveying maps (such as θ).
Note that the rapid developments in GPS technology also enable acquisition of this
data. Further, more detailed discussion is available in the monograph by Muneer
et al (2017).

5.6 Supporting Software

To run the VEDEC simulation, one has to first obtain angle of gradient and distance, θ and d for the route at frequent intervals of distance traveled. Any good GPS
tracking system will record and download such information. 'Mapometer' is an
online altimeter that provides an accurate measure of topography. The validation of

Table 5.2 Validation of the supporting software program Mapometer (www.mapometer.com)

	Measured values		Mapometer values		Error	
Experiment	Distance, m	Altitude, m	Distance, m	Altitude, m	Distance (%)	Altitude (%)
1	1276.5	73.8	1280	78	0.3	5.7
2	421.6	33.4	430	33	2.0	1.2

this software is illustrated in Table 5.2. Alternative methods to obtain the angle of altitude and distance traveled are by using an on-board altimeter or by using topography maps if available.

5.7 Validation of Simulations

To validate the model, energy simulated is compared with energy measured by the test vehicle's on-board display of energy consumption data for traction energy, recovered energy due to braking and climate control at the end of each trip. During periods of hard braking when the driver applies force to the brakes, friction between the brake pads and tire will cause the vehicle to decelerate. Not all energy will be regenerated by the braking system as frictional braking will account for the majority of braking. Therefore, there will be an increased error in the simulation when calculating recovered energy as efficiency will vary depending on driver's braking style.

Results for 11 routes undertaken within the City of Edinburgh were analyzed by using the experimental vehicle (Renault, Zoe) to validate the simulation as shown in Table 5.3. The test vehicle's built-in algorithm that computes energy consumption displays to only one decimal place.

Note that File Ex05-01.xlsm also provides overall statistics of the simulation for the journey thus:

E used: The energy demand of the vehicle.

E regen: The regenerated energy captured by the vehicle.

E tot: The net energy used by the vehicle.

Average speed: The vehicle average speed during the entire journey.

Total distance: The total distance traveled by the vehicle in meters.

Table 5.3 Test runs undertaken by Renault Zoe (Edinburgh based)

No	Route	Average Speed (mph)	Simulation		Experiment		Computational Accuracy	
			Energy used[a] (kWh)	Energy regerated[b] (kWh)	Energy used (kWh)	Energy regenerated (kWh)	Traction program error (%)	Regeneration program error (%)
1	Morningside-Leith	17	1.14	0.36	1.1	0.3	4	21
2	Leith-Morningside	16	1.41	0.23	1.5	0.2	−6	15
3	Home-Sighthill	25	2.12	0.47	2.2	0.5	−3	−6
4	Sighthill-Home	36	3.32	0.48	3	0.7	11	−32
5	Home-Greens	23	1.33	0.33	1.4	0.4	−5	−18
6	Greens-Home	20	1.49	0.33	1.4	0.4	6	−18
7	Home-Costco	24	1.67	0.22	1.7	0.2	−2	12
8	Costco-ESR	26	1	0.37	0.9	0.4	12	−7
9	Napier-Sighthill	25	1.21	0.31	1.1	0.3	10	3
10	Sighthill-Napier	25	1.33	0.23	1.3	0.2	2	14
11	Napier-Dalkeith	26	2.96	0.8	2.9	0.9	2	−11

5.8   MS Excel-VBA Functions in VEDEC Simulation Software

The two functions below are part of the VEDEC simulation software (File: Ex05-01.xlsm refers).

The above function identifies one of the nine possible modes of travel: 'Accelerating Up-gradient', 'Accelerating Down-gradient', 'Accelerating Level', 'Decelerating Up-gradient', 'Decelerating Down-gradient', 'Decelerating Level', 'Cruise Up-gradient', 'Cruise Down-gradient', or 'Cruise Level'. The identified mode is then recorded within the spreadsheet.

Table 5.4   Function DriveMode(speed1, speed2, alt1, alt2)

```
Function distpres(time1, time2, speed, distance)
speedmps = 0.44704 * speed
distpres = distance + speedmps * (time2 - time1)
End Function
```

The above function calculates the distance travelled by the vehicle between two time events, provided the vehicle speed is given in miles per hour.

```
Function accelerate(speed1, speed2, time1, time2)
accelerate = 0.44704 * (speed2 - speed1) / (time2 - time1)
End Function
```

The above function computes vehicle acceleration.

```
Function DriveMode(speed1, speed2, alt1, alt2)
If (speed2 > speed1) Then
If (alt2 > alt1) Then
DriveMode = "Accelerating Up-gradient"
ElseIf (alt2 < alt1) Then
DriveMode = "Accelerating Down-gradient"
Else
DriveMode = "Accelerating Level"
End If
ElseIf (speed2 < speed1) Then
If (alt2 > alt1) Then
DriveMode = "Decelerating Up-gradient"
ElseIf (alt2 < alt1) Then
DriveMode = "Decelerating Down-gradient"
Else
DriveMode = "Decelerating Level"
End If
Else
If (alt2 > alt1) Then
DriveMode = "Cruise Up-gradient"
ElseIf (alt2 < alt1) Then
DriveMode = "Cruise Down-gradient"
Else
DriveMode = "Cruise Level"
End If
End If
End Function
```

The file Ex05-01.xlsm has three worksheets. The sheet 'Notes' is non-interactive and merely provides the governing equation. Start off by providing data in the yellow colored worksheets 'Data1' then 'Main'. You will obtain results in the light blue part of the two sheets. The code uses user-defined functions, and hence, there is no need to execute on the part of the user. The code will update calculations automatically.

5.9 Conclusion

Electric vehicles are rapidly being introduced within any given country's fleet. This chapter presents the basic dynamical analysis of the energy consumed by the latter vehicles during traction mode and regained while braking. Measured data was presented for one such vehicle and used for validating the VBA code that was presented in this chapter for analyzing the above energy transfers.

References

K. Holmberg, P. Anderson, A. Erdemir, Global energy consumption due to friction in passenger cars. Tribol. Int. **47**, 221–234 (2012)

T. Muneer, R. Milligan, I. Smith, A. Doyle, M. Pozuelo, M. Knez, Energetic, environmental and economic performance of electric vehicles: experimental evaluation. Transp. Res. Part D: Transport Environ. **35**, 40–61 (2015)

T. Muneer, M. Kolhe, A. Doyle, *Electric Vehicles: Prospects and Challenges* (Elsevier, Amsterdam, 2017)

B. Randall, Electric cars—are they really green? in *Proceedings of IMechE Conference on 'Low-Carbon Vehicles 2009'*, London (2009), pp. 17–25

Thermodynamics

6

Thermodynamics is that branch of physics that deals with the laws that govern the inter-relationship between heat, mechanical work, and energy. Properties of substances such as temperature, pressure, density, internal energy, enthalpy, entropy, and specific heat are of fundamental importance in the study of thermodynamics. Another definition of thermodynamics is the study of enthalpy and entropy.

The four laws of thermodynamics—zeroth, first, second, and the third law—are of profound significance in the understanding and analysis of very many physical processes.

Lord Kelvin William Thomson was born in 1824. In the year 1846, he got the Chair of Natural Philosophy at the University of Glasgow. In the year 1849, he published the treatise of his work that included the very first use of the word thermodynamics.

In this chapter, we shall explore the analysis of a few thermal systems that will be of use to building services, energy, chemical, mechanical, and process engineers. A good reference for analyzing systems that vary from simple to complex in their content of vigor is W. F. Stoecker, design of thermal systems, third edition, McGraw-Hill, 1989.

Example 6.1 Charging of an evacuated container with water/steam substance

Problem definition

In dealing with pure substances such as water/steam, it has to be borne in mind that the ideal gas laws are inapplicable. The pressure–temperature density relationships are much more involved. The calculation of other properties such as enthalpy, entropy, internal energy, and dryness fraction is also quite involved. The International Association for Properties of Water and

---

Supplementary Information The online version contains supplementary material available at https://doi.org/10.1007/978-3-030-94085-0_6.

T. Muneer and S. Ivanova, *Excel-VBA*,
https://doi.org/10.1007/978-3-030-94085-0_6

Steam (IAPWS) oversees the mathematical formulations for the thermodynamic properties mentioned above.

The IAPWS website provides all the internationally agreed forward and backward formulations the thermodynamic properties of the substance and all material may be accessed via this website:

http://www.iapws.org/newform.html

The above website provides downloadable calculators for thermodynamic properties that are available for a small fees.

Thermodynamic and thermo-physical properties of water/steam

Figure 6.1 shows the temperature-entropy chart for water/steam substance. Any computer code that is used for obtaining thermodynamic properties such as pressure, temperature, density, dryness fraction for the saturation mixture region, enthalpy, entropy, and internal energy has to conform to the latest standards. The chart shown in Fig. 6.1 shows six of the above seven properties. The Association defines the thermodynamic properties of the fluid phases of water and steam over a wide range of temperature and pressure. IF97 identifies four main regions in the temperature-enthalpy diagram for the fluid phases of water and steam which correspond to the major states of the substance. Table 6.1 identifies the regions, the corresponding states of the substance, and the fundamental equations that can be used to model substance behavior in each case.

In order to reduce computing effort, IF97 also provides a series of 'backward equations' of the form T (P, h) for regions 2 and 3 for the calculation of thermodynamic properties as functions of pressure and enthalpy. The regions are shown in Fig. 6.2. These equations are numerically consistent with the basic ones and allow calculation of properties without the need for iteration. The backward equations effectively fragment the regions as shown in Fig. 6.2. The substance also exhibits specific behavior at the boundary between regions 2 and 3 (shaded area in Fig. 6.2), and IF97 provides an auxiliary equation for the calculation of properties in that area known as the B23-equation. The backward equations and the B23-equation are all implemented in the VBA code which takes starting pressure and enthalpy as its initial parameters.

Presently, a VBA code (Ex06-01.xlsm) is presented that will allow the user to obtain other properties for the water/steam substance provided two independent thermodynamic properties are provided as input. The operation of the VBA code is based primarily on the Revised Release on the IAPWS Industrial Formulation 1997 for the Thermodynamic Properties of Water and Steam (IF97). That formulation is provided at the Association's website: http://www.iapws.org/newform.html.

The VBA program code can be found within the MS Excel/VBA file Ex06-02.xlsm. The output properties generated by the code are shown in Table 6.2. Figure 6.3 shows the information flow diagram for the code under discussion.

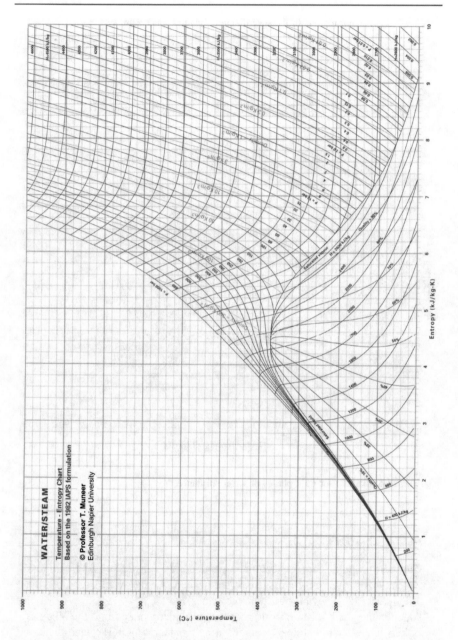

Fig. 6.1 Temperature = entropy chart for water/steam substance

Table 6.1 Four major regions in the state diagram for fluid water substance

Region	Phase	Abbreviation	Equation
1	Compressed liquid	CL	Gibbs free energy: g(p, T)
2	Superheated vapor	SUPV	Gibbs free energy: g(p, T)
3	Supercritical liquid	SCR	Helmholtz free energy: f(ρ, T)
4	Saturated mixture	SATM	Saturation pressure equation: $P_s(T)$

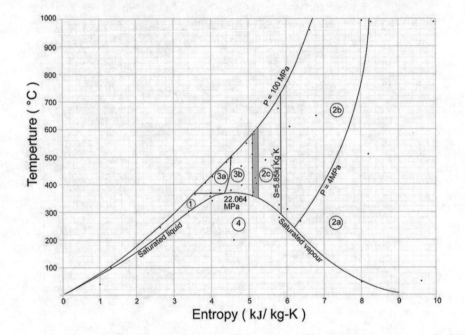

Fig. 6.2 State diagram for fluid phases of water substance

Thermo-physical properties

Transport properties are also required when energy transfer and frictional loss calculations for thermo-fluid systems are needed. As shown in Table 6.2, the code returns the relevant transport properties.

The reference data for transport properties was incorporated into the VBA as lookup tables. While Excel provides a convenient LOOKUP facility as a standard function, it only provides for one-dimensional search and cannot provide automatic interpolation between reference points. The VBA therefore contains an independent VBA routine to provide one- or two-dimensional interpolation for water transport properties as appropriate.

Table 6.2 VBA program output

Symbol	Description
p	Pressure (bar)
h	Enthalpy of water/steam substance (kJ/kg)
T	Temperature (K)
T_{sat}	Saturation temperature (K)
h_f	Specific enthalpy of saturated liquid (kJ/kg)
s_f	Specific entropy of saturated liquid (kJ/kg K)
ρ_f	Density of liquid (kg/m^3)
h_g	Specific enthalpy of saturated vapor (kJ/kg)
s_g	Specific entropy of saturated vapor (kJ/kg K)
ρ_g	Density of vapor (kg/m^3)
state	State of the substance
x	Dryness fraction
h_b	Enthalpy for boundary (kJ/kg)
Region	Sub-region of the state chart corresponding to the particular combination of characteristics
Pr	Prandtl number
λ	Thermal conductivity (W/m K)
μ	Kinematic viscosity (m^2/s)
s	Specific entropy (kJ/kg K)
ρ	Density (kg/m^3)

The interpolation requirements for water transport properties are dependent on the state of the substance. In the saturated mixture state, properties are only dependent on temperature, and a one-dimensional search is required; as compressed liquid or superheated vapor properties depend on both temperature and pressure, a two-dimensional search must be used which is presently used. In the present example, we shall deal with filling of an evacuated container with steam in a Uniform State Uniform Flow (USUF) process. In such process, the thermodynamic properties at a given point within the control volume have the same value at any instance.

Refer to Fig. 6.4. Once the flow valve is opened, the high pressure steam from the supply line will fill the vessel until the pressure inside it is in equilibrium with the supply line which has, respectively, a pressure and temperature of 0.6 MPa and 360 °C. Calculate the final temperature of steam in the charged vessel given that it is effectively insulated.

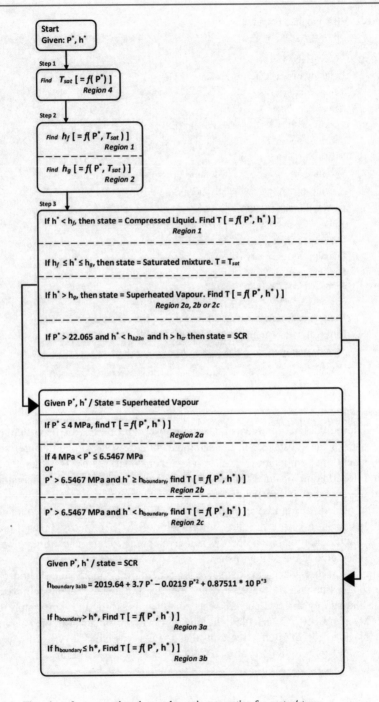

Fig. 6.3 Flowchart for computing thermodynamic properties for water/steam

Fig. 6.4 An evacuated vessel connected to a high pressure steam line

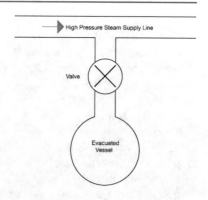

Solution

From mass continuity equation, we may write for steam,

$$m_{in} = m_{final}$$

From first law of thermodynamics, we write,

$$m_{in} \cdot h_{in} + Q_{in-final} = W_{in-final} + m_{final} \cdot u_{final}$$

Since $Q_{in\text{-}final}$ and $W_{in\text{-}final}$ are both nil, the above relationship reduces to Eq. 6.1.1,

$$h_{in} = u_{final} \tag{6.1.1}$$

Note that using published steam tables, one may find h_{in} for the given properties. An extract of such tabulated properties is shown in Table 6.3. By interpolation, we get the value of enthalpy, h at 0.6 MPa and 360 °C as 3187 kJ/kg. One may also get any thermodynamic property for the latter combination of pressure and temperature from IAPS-based formulation, as shall be demonstrated here.

Launch the Excel-VBA file 'Ex06-01.xlsm' and after clicking ALT + F11 keys launch the VBA code. Even a cursory examination will reveal the

Table 6.3 Properties for superheated water/steam at $p = 0.6$ MPa (t-saturation = 158.8 °C)

t, °C	200	300	400	500	600	700
ρ, kg/m³	2.8409	2.3020	1.9470	1.6898	1.4934	1.3385
h, kJ/kg	2849.7	3062.3	3270.6	3482.7	3700.7	3925.1
s, kJ/kg-K	6.9662	7.374	7.709	8.0027	8.2678	8.5111

complexity of the code which utilizes hundreds of coefficients. There are
many engineering applications that require other thermodynamic properties
from a pressure-enthalpy implicit base. IAPS have provided formulations that
are pressure–temperature implicit which are equally involved. Presently, we
shall use the pressure-enthalpy implicit VBA code which is part of the
Ex06-01.xlsm file.

There are seven worksheets within the Ex06-01.xlsm file. The only sheet you
need to focus on is the 'Input' worksheet. Columns A and B contain the input
parameters, pressure and enthalpy, which trigger the code to generate a large
number of output variables which are provided in columns C through to V,
the latter containing the value of u, internal energy. Refer to Eq. 6.1.1. The
value of u we are seeking is 3187 kJ/kg. Note that $h = u + p/\rho$, and hence,
we try for a solution with a value of $h = 3400$ kJ/kg which results in
$u = 3063$ kJ/kg. We get the solution via building a grid from cells 2 to 25 in
which h is increased in increments of 10. We find that $h = 3560$ kJ/kg provides the solution. The temperature at this state point is 535.4 °C (column O).

Discussion

Water/steam is one of the most used substances in engineering practice.
A useful code is provided in this example that provides most of the useful
thermodynamic properties that are obtained from the latest IAPS formulation.
The coding is very complex and it will take a while for the reader to get
accustomed to this VBA code.

A grid-based approach was used to obtain the solution to a set of complex
nonlinear equations. If a more forthright solution is sought, then the reader may use
the bisection method introduced in Chap. 3. The reader is being encouraged to do
just that via an exercise that follows.

Example 6.2 Power output from a steam turbine

Problem definition

In the previous Example 6.1, we dealt with the Uniform State Uniform Flow
(USUF) process. In engineering design and practice, one encounters frequently what is known as the Steady State Steady Flow (SSSF) process. The
latter shall be the subject of this example.

A steam turbine takes in steam at a pressure of 40 bar, 400 °C (inlet condition, i) and exhausts steam at 4 bar pressure (exit condition, e) after the steam

has undergone expansion in an isentropic (constant entropy) process. Obtain the power output of the turbine for a steam flow rate of 1 kg/s.

Solution

On the basis of the first and second law of thermodynamics, we may write Eqs. 6.2.1 and 6.2.2,

$$\text{Power output, } \dot{W} = \dot{m}(h_i - h_e) \qquad (6.2.1)$$

$$s_e = s_i \qquad (6.2.2)$$

The solution strategy will be to use the given inlet conditions of pressure and temperature of 40 bar and 400 °C, respectively, to obtain enthalpy and entropy for the inlet conditions, h_i and s_i. Furthermore, since $s_e = s_i$ and $p_e = 4$ bar, the enthalpy h_e can then be obtained. Equation 6.2.1 will then provide us with our desired result.

Launch the file Ex06-02a.xlsm which provides the first part of the solution: obtaining h_i for the given inlet conditions of pressure and temperature of 40 bar and 400 °C. You only need to use the 'Input' worksheet. Note that the software code provided is implicit in pressure and enthalpy; hence, we will use an exhaustive search procedure to get our solution. This is done by fixing the value of $p = 4$ MPa in cells A2 through to A47 and providing trial enthalpy values of 160- through to 4600 kJ/kg in column B. We are thus moving along the 4 MPa isobar and calculating thermodynamic properties as we sweep through the range of enthalpy values. Note that before you do anything you need to delete all data from columns 'C' through to 'U'.

Launch the macro dialog box by simultaneously clicking ALT + F8 keys. Once the dialog box opens, you will see the single VBA code named 'WATERSTEAM'. Select this code and then click the Run icon. The 'Input' worksheet will now be populated with all of the required (and more) thermodynamic properties. Examine the contents of the cells O33 and O34 that now contain the temperature values of 393.9 and 436.6 °C. Our given temperature is 400 °C which is in between the latter values, and hence, a quick interpolation shall provide us with our desired enthalpy and entropy for the inlet conditions of steam (see cells W33 and W34). We thus have:

$$h_i = 3214.23 \text{ kJ/kg and } s_i = 6.7704 \text{ kJ/kg-K}$$

We are now in a position to obtain h_e as we know s_e (=6.7704) and p_e (= 0.4 MPa), and we shall use the code given in Ex06-02b.xlsm for that task. Once again, we shall fix the pressure value at 0.4 MPa in column A and provide trial values of enthalpy, h in column B. A quick look at Fig. 6.1 will direct us to select the range of enthalpy values.

Launch the macro dialog box of the file Ex06-02b.xlsm by simultaneously clicking ALT + F8 keys. Once the dialog box opens, you will see the single VBA code named 'WATERSTEAM'. Select this code and then click the Run icon. The 'Input' worksheet will now be populated with all of the required (and more) thermodynamic properties. Examine the contents of cells T60 and T61 that now contain the entropy values of 6.756113 and 6.780108 kJ/kg-K. These values are bracketing the given value of s_e = 6.7704 kJ/kg-K, and hence by interpolation, we get h_e = 2685.95 kJ/kg in cell W61.

From Eq. (6.2.1) we get, power output = (3214.23 − 2685.95) = 528.3 kW.

Discussion

We have used the IAPWS formulation for water/steam substance to build a VBA code that can provide a powerful tool for analyzing problems related to thermodynamics. We have taken as sample problems that dealt with Uniform State Uniform Flow (USUF) and Steady State Steady Flow (SSSF) processes. The technique adopted was exhaustive search, and this is a most robust technique as the selection of trial values should be undertaken with great care after examining the thermodynamic envelope for the pure substance under review. We shall extend this analysis by using the bisection method which was the subject of Example 3.2.

Note that a fuller discussion on large-scale solar thermal power generation from open systems is provided in the following two references: Aldali et al. (2009) and Aldali et al. (2012).

Example 6.3 Obtaining thermodynamic properties iteratively via use of bisection method

Problem definition

Calculate thermodynamic properties for the water/steam substance for a given pressure of 4 MPa and a temperature of 400 °C. Use bisection method to solve the pressure-enthalpy implicit formulation.

In the above two examples, it was shown that the presently developed VBA code which is pressure and enthalpy implicit may be used to compute thermodynamic properties when other input properties are provided such as pressure and temperature, pressure and entropy, or pressure and internal energy. We can use the bisection method instead of exhaustive search for this task. The merits and demerits of the two methods shall also be discussed.

Solution

Launch the file Ex06-03.xlsm and examine the first 42 lines of code provided (see Table 6.4). You will be able to review that code by pressing keys Alt + F11.

The code may be divided into five sections as shown above. The first part merely provides the input values of pressure and temperature, followed by the second part that executes the bisection algorithm. The third part sets the tolerance limit for the computation followed by executing a command to stop the execution if the routine undertakes more than a thousand iterations. The last part provides the output.

The 'RESULTS' worksheet provides the value of enthalpy that we seek in cell B18 as h = 3214.39 kJ/kg. The required number of iterations were 18 as shown in cell B19. The last row of worksheet 'Input' presents other converged values of thermodynamic properties.

Discussion

An iterative algorithm, rather than the previously used exhaustive procedure, was demonstrated here to obtain thermodynamic properties of a pure substance. Care should, however, be taken when providing initial, seed values for the given properties by carefully studying the chart for thermodynamic properties as shown in Fig. 6.1.

6.1 Conclusion

A VBA code was presented in this chapter that will allow the user to obtain thermodynamic for the water/steam substance. The code is based on the mathematical formulations developed by the International Association for Properties of Water and Steam (IAPWS) which oversees all such developments. Additionally, the code also delivers transport properties for water and steam.

The applications introduced in this chapter included Steady State Steady Flow (SSSF) and Uniform State Uniform Flow (USUF) processes.

Exercises

E6.1 Refer to Example 6.1 in which the solution for internal energy, u, was obtained through a grid-based approach. Use the bisection method to obtain the above solution.

Table 6.4 VBA code for the main part of the program for water/steam properties

```
w = 2
P = 4
Tgiven = 400
X1 = 160
X2 = 4600
niter = 0
Line2:
X = X1
h = X
GoSub 101
Y1 = (T1 - Tgiven)
X = X2
h = X
GoSub 101
Y2 = (T1 - Tgiven)
XM = 0.5 * (X1 + X2)
X = XM
h = X
GoSub 101
YM = (T1 - Tgiven)
' TOLERANCE FOR FUNCTION SET AT 1E-8
If (Abs(YM) < 0.001) Then
GoTo Line1
ElseIf (Y1 * YM) > 0 Then
X1 = XM
Else
X2 = XM
End If
If (niter > 1000) Then
GoTo Line1
Else
w = w + 1
w = 2
niter = niter + 1
Sheets("RESULTS").Cells(niter + 1, 4) = X1
Sheets("RESULTS").Cells(niter + 1, 5) = X2
Sheets("RESULTS").Cells(niter + 1, 6) = YM
GoTo Line2
End If
Line1:
Sheets("RESULTS").Cells(18, 2) = XM
Sheets("RESULTS").Cells(19, 2) = niter
End
```

E6.2 Refer to Example 6.1 once again. Note that for 0.6 MPa and 360 °C the value of h was obtained as 3187 kJ/kg via Table 6.3. Set up the VBA code provided in Ex06-02.xlsm so that the temperature of 360 °C is obtained for the input parameter combination of $p = 0.6$ MPa and $h = 3187$ kJ/kg.

E6.3 Refer to Table 6.3. Use the VBA code for water/steam substance that is presently provided to obtain solution for density, entropy, and temperature for various combinations of pressure and enthalpy and compare with the values reported in the latter table of thermodynamic properties.

References

Y. Aldali, D. Henderson, T. Muneer, Prospects for large-scale solar thermal electricity generation from the Libyan desert: technical feasibility, in *Es2009: Proceedings of the ASME 3rd International Conference on Energy Sustainability*, vol. 2 (2009)

Y. Aldali, B. Davison, T. Muneer, D. Henderson, Modelling the behaviour of a 50 MW direct steam generation plant for Southern Libya based on the thermodynamic and thermo-physical properties of water substance. J. Solar Energy Eng.-Trans. ASME **134**(4) (2012)

Solar Energy

7

Solar energy has come of age. From its humble beginnings in the early 1900s when water was being heated in plastic 'pillows' by solar energy in Florida and Japan, mega-scale power plants are now in operation that are based on thermodynamic cycle or photovoltaic modules. This chapter presents examples that deal with solar geometry, equation of time, ambient temperature modeling, and photovoltaic design.

Example 7.1 Solar geometry: Kerala Temple

An almanac provides information regarding tide times, and other data such as the rising and setting times of the Sun, Moon, and planets of our solar system as well as dates of eclipses. In the present example, we shall explore algorithms that will enable us to understand and obtain solar geometry. We will explore this topic with the help of an ancient architectural structure that used solar design.

The Padmanabhaswamy temple in Kerala state, India, is a Hindu place of worship and meditation that has existed for around 1500 years. The structure is to be found in the city of Thiruvananthapuram which is the new name of Trivandrum, a major city and capital of the latter state. The tower structure that is based on solar design is called '*gopuram*' and is an extension which was built in the year 1566. The latter structure is 30 m high and is divided into seven tiers.

The main solar feature of the *gopuram* is that on the two days of equinoxes the Sun will set in a plane that is exactly perpendicular to the tower. The tower has five windows and close to sunset it will appear, in turn, in direct view of each of those windows with five-minute intervals. There are a number of videos available on the YouTube, one of which has the web link provided below:
https://www.youtube.com/watch?v=NaLTdL9CEzQ
Problem definition

Supplementary Information The online version contains supplementary material available at https://doi.org/10.1007/978-3-030-94085-0_7.

You are asked to come up with a solar architecture design that would provide
details regarding the height of each of the above-mentioned five windows and
other details such as the window width and the slope of the sill. The viewers
will position themselves in a line that is perpendicular to the tower at a
distance of 150 m.

Solution

Launch the Excel/VBA file Ex07-01.xlsm and explore the data that is to be
provided on the left-hand side of the dividing red line and cells E7 and F7 that
are respectively provided to key-in the time for which solar geometry is
needed. The output is provided in green-colored cells and contains information related to solar elevation, solar azimuth, and incidence angle of the Sun
on the surface whose geometry is provided in cells B9 and B10. Cell G7
provides Apparent Solar Time (AST) corresponding to the clock time key-in
cells E7 and F7. Figure 7.1 should be referred to gain an understanding of the
geometrical parameters discussed above. Further information is provided in
Muneer (2004).

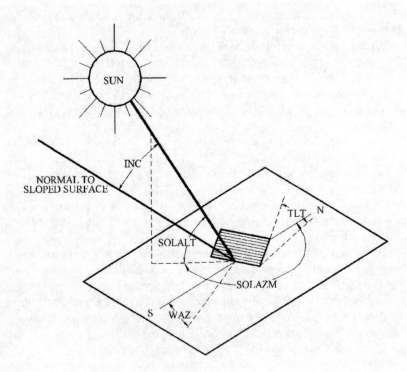

Fig. 7.1 Solar geometry

By manipulation of data in cells E7 and F7 of the referred Excel workbook, we find the closest time for sunset is 18:30. If we then compute the solar altitude from a time that is ten minutes prior to sunset and work backward, we can prepare Table 7.1 that provides the solar elevation and the height above ground level for the center of the windows to be erected.

Table 7.1 Solar elevation and window height to capture Sun at Padmanabhaswamy temple

Hour	Minute	Solar altitude (degree)	Window height (m)
18	0	7.3	19.1
18	5	6.1	15.9
18	10	4.9	12.8
18	15	3.6	9.4
18	20	2.4	6.3

Note that the angle subtended by the Sun to an observer on planet Earth is given in literature as 0.52° which may easily be checked from Sun's distance from Earth and Sun's diameter which respectively are, 149.6- and 1.3927 billion meter. The ratio of the latter to former figure is 0.009309 rad which translates to 0.53°. The latter value is close to the literature-cited figure of 0.52°.

The appropriate width of the window for the Sun to be completely visible from a distance of 150 m can thus be worked out from the Sun's subtended angle of 0.52° as 1.4 m.

Discussion

The Excel/VBA file Ex06-01.xlsm is a powerful tool to obtain very accurate solar geometry, and with the growing interest and application of solar energy, it may be put to good use. The accuracy of the computations is high and has been rigorously checked for the past three decades. For further information, the reader is directed to the following publications which will provide much more in-depth discussion: Muneer et al. (2000) and Muneer (2004).

Example 7.2 Equation of time

The simplest definition of the 'equation of time' is that it is the difference between a sundial time and an electronic or mechanical clock which runs at a constant speed.

A poem by Tad Dunne goes like this [https://en.wikipedia.org/wiki/Equation_of_time]:

On September one, trust the Sun

Come Halloween, subtract sixteen

On Christmas Day, you're OK

For your Valentine true, add a dozen and two

The mid of month four, add no more

At the mid of May, take four away

On June fourteen, don't add a bean

When August begins, add seven little mins

The rest is easy: for any date, all you do is interpolate

Problem definition

Among the other output variables, the Excel/VBA file Ex06-01.xlsm provides the value of the Equation of Time (EoT), it being presently defined as the difference between the Apparent Solar Time, as recorded by a sun dial, and an electronic or mechanical clock located at the standard time meridian.

Some people define the EoT the other way round, i.e. difference between a clock time and sundial time as is the case with the poem narrated above.

Use file Ex07-01.xlsm to validate Tad Dunne's poem.

Solution

Launch the Excel/VBA file Ex07-01.xlsm and provide the dates mentioned in the poem, in each case obtaining the value of EoT. In this manner, Table 7.2 may be constructed.

Table 7.2 Validation of Tad Dunne poem on 'equation of time'

Date	EoT (min)
September 1	0.0
October 31	16.4
December 25	−0.1
February 14	−14.2
April 15	0.0
May 15	3.6
June 14	−0.3
August 1	−6.4

Note that the solar declination and equation of time formulations used within the file Ex07-01.xlsm are based on the work of Yallop of Greenwich

Observatory, England. This high accuracy formulation is valid for the period
1980–2050. Further information is provided in T. Muneer, solar radiation and
daylight models, Elsevier, 2004.

Discussion

Generally speaking, the claims made in the poem are valid, the only slight
difference being for August 1. Another thing one has to bear in mind is the
way EoT is defined. In the case of Tad Dunne's poem, the equation that
relates sundial to clock time is the difference between the latter and the former
time system.

Example 7.3 Day-time mean ambient temperature

The oldest temperature records for England go back right up to seventeenth
century. Newspapers have been publishing the daily maximum and minimum
temperature for principal locations for over hundred years. Now, with the
advent of smart phones, weather data is on everybody's fingertips which also
provide hourly temperatures. Those hourly values may easily be computed if
one has access to daily maximum and minimum temperature. That will be the
subject of the present example.

Temperature data is of importance in building services design and operation as it dictates the energy use for heating and cooling. Day-time mean
temperature is needed in applications such as solar water heating and PV
design as the latter parameter affects the system performance. The main item
that this example deals with is how to use the daily minimum/maximum
temperature data to obtain an average value of temperature for daylight hours.

A large body of research has been carried out by two premier professional
engineering societies: American Society for Heating, Refrigerating and
Air-conditioning Engineers (ASHRAE) based in Atlanta, and the Chartered
Institution of Building Services Engineers, London. Using the work of
ASHRAE, it is possible to find hour-by-hour temperature from daily maximum and minimum temperature. Refer to Table 7.3 and Eq. 7.3.1 which
jointly enable such estimation.

The dimensionless parameter, Z, introduced in Table 7.3 is defined as the
ratio of hourly temperature elevation to daily range. Thus, using Eq. 7.3.1,

$$Z = \{(Th - Tmin)/(Tmax - Tmin)\} \qquad (7.3.1)$$

Th is the temperature for any given hour (the desired quantity) and Tmin
and Tmax are the daily minimum and maximum temperatures that are provided. File Ex07-02.xlsm enables such calculation. Note that only the sheet
'Temperature' is to be used as the other sheet contains algorithm-related
information. Cell G1 contains the day-time average temperature, presently

Table 7.3 ASHRAE model for diurnal temperature swing

Hour	Z	Hour	Z
1	0.12	13	0.95
2	0.08	14	1.00
3	0.05	15	1.00
4	0.02	16	0.94
5	0.00	17	0.86
6	0.02	18	0.76
7	0.09	19	0.61
8	0.26	20	0.50
9	0.45	21	0.41
10	0.62	22	0.32
11	0.77	23	0.25
12	0.87	24	0.18

Table 7.4 Calculation of hourly temperature from diurnal temperature swing

```
Sub TempCalcs()
    tmax = Sheets("Temperatue").Cells(2, 1).Value
    tmin = Sheets("Temperatue").Cells(2, 2).Value
    For mnhour = 1 To 24
        FracTemp = Sheets("Data").Cells(mnhour, 1).Value
        Tair = tmin + FracTemp * (tmax - tmin)
        Sheets("Temperatue").Cells(1 + mnhour, 4) = Tair
    Next mnhour
End Sub
```

using data from cell range D7:D19 which corresponds to the time period 0600 to 1800 h. For day length that is shorter or longer, the latter cell range has to be altered. Table 7.4 provides the VBA code.

Further validation information on the hourly temperature model is provided in Gago et al. (2010).

Example 7.4 Solar PV design

Jeffrey et al (2015) have presented a detailed calculation procedure that leads to obtaining the power output (Watt) of solar PV modules.

The procedure requires in the main the following four steps:

a- Provide the required data: horizontal global (total) and diffuse irradiation (W/m²), ambient temperature, time of the day, surface geometry, and characteristics for the PV module.
b- Compute slope irradiation on the PV module which has three components: beam, sky-diffuse, and ground-reflected radiation.
c- Compute module efficiency from its characteristics.
d- Compute the power output of the module.

Jeffrey et al. (2015) have also presented measured data for each of the above calculation steps with the view to validate the entire procedure. Steps 'b' and 'c' are undertaken, respectively, using the algorithms presented by Muneer (2004) and this website: https://www.homerenergy.com/products/software. html

The entire procedure is demonstrated via present example.

Problem definition

Data for ambient temperature, horizontal global, and diffuse irradiation for Edinburgh, Scotland (Latitude = 56° North, Longitude = 3.2° West), for August 12, 1993, is provided in the file Ex07-03.xlsm. A PV module manufacturer has provided the following data: power output under STC conditions: 550 W, temperature coefficient of P_{max} (α_p): −0.35%/°C, T_{NOCT}: 45 °C, module efficiency: 21.5%.

Calculate the power output for the module given its orientation and tilt from horizontal are 20° West of South and 35°. Note that within the present workbook the aspect is measured from North. Hence, the aspect for East, South, and West will respectively be 90°, 180°, and 270°.

Note:

STC: Air Mass 1.5, 1000 W/m², 25 °C.

Nominal Operating Conditions: Air Mass 1.5, 800 W/m², 20 °C, 1 m/s.

Solution

File Ex07-03.xlsm displays the solution to the present problem. All solar radiation calculations are performed as per algorithms presented in T. Muneer, solar radiation and daylight models, Elsevier, 2004 and the cell temperature obtained from Eq. 7.4.1a,

Model for cell temperature,

$$T_c = T_a + (T_{c,noct} - T_{a,noct}) \frac{G_{slope}}{G_{noct}} [1 - \eta_{stc}/\tau\alpha)] \qquad (7.4.1a)$$

Note that $T_{a,noct} = 20$ °C. $G_{noct} = 800$ W/m², $\tau\sigma = 0.92$.

PV cell efficiency is obtained from Eq. 7.4.1b,

$$\eta_{cell} = \eta_{noct}[1 + \alpha_p(T_c - T_{c,noct})] \qquad (7.4.1b)$$

Discussion

File 07-03.xlsm contains a large number of VBA functions that cover solar geometry as well as solar irradiance calculations. The computations then proceed to provide cell temperature, cell efficiency, and module output. You need to study the progression of the computations carefully using the material presented in the above two references that were cited along with Eq. 7.4.1. Solar energy is gaining serious ground in very many countries, and the present code will be of help for simple PV design tasks.

7.1 Conclusion

Solar energy and its applications is now a very much established and worthy science. This chapter presented applications related to the basics of solar time, solar geometry, and elementary solar photovoltaic design. Also included were algorithms that enable decomposition of daily-to-hourly ambient temperature as that information is needed in solar thermal and PV design. The chapter included applications that the reader may find of interest.

References

E. Gago, S. Etxebarria, Y. Tham, Y. Aldali, T. Muneer, Inter-relationship between mean-daily irradiation and temperature, and decomposition models for hourly irradiation and temperature. Int. J. Low-Carbon Technol. (Oxford University Press, 2010). https://doi.org/10.1093/ijlct/ctq0392010

T. Muneer, N. Abodahab, G. Weir, J. Kubie, *Windows in Buildings (with CD)* (Architectural Press, 2000)

T. Muneer, *Solar Radiation and Daylight Models* (Elsevier, Oxford, 2004) (with companion suite of computer programs)

M. Jeffrey, I. Kelly, T. Muneer, I. Smith, Evaluation of solar modelling techniques through experiment on a 627 kWp photo-voltaic solar power plant at Edinburgh College—Midlothian Campus, Scotland. J. Renew. Sustain. Energy **7**, 033128 (2015)

Heat Transfer and View Factors

8

8.1 Heat Transfer—Introduction

There are several areas of applicability of radiation heat transfer to the cases discussed in this chapter:

i. Solar air temperature and cooling load of the building due to energy exchange from the ground and adjacent surfaces of the nearby building;
ii. Determining the contribution of the reflected solar radiation from the Earth and nearby sites to the energy balance of solar thermal collectors and photovoltaic modules.

The purpose of this chapter is to present procedures for the interfacial exchange of radiant energy.

8.1.1 Heat Transfer and View Factors

The thermal radiation exchange between any two black surfaces is given by Eq. (8.1a):

$$Q_{1-2} = \sigma(T_1^4 - T_2^4)A_1F_{1-2} = \sigma(T_2^4 - T_1^4)A_2F_{2-1} \qquad (8.1a)$$

The term $F_{1\text{-}2}$ is known as 'configuration factor' (*CF*) (Howell 1982a) within heat transfer terminology. The other names for this term are 'angle factor', 'view factor', 'geometry factor', or 'shape factor'. In this chapter, we will use the most popular term view factor (VF).

Supplementary Information The online version contains supplementary material available at https://doi.org/10.1007/978-3-030-94085-0_8.

The view factor $F_{1\text{-}2}$ between any two elemental surfaces dA_1 and dA_2 as those shown in Fig. 8.1 is given as Eq. (8.1b):

$$F_{1-2} = \frac{1}{A_1} \int_{A_1} \int_{A_2} \frac{\cos \Phi_1 \cos \Phi_2}{\pi R^2} dA_2 dA_1 \qquad (8.1\text{b})$$

where A_1 and A_2 are the faces of both surfaces, R is the distance between both differential elements dA_1 and dA_2; ϕ_1 and ϕ_2 are the angles between the normal vectors to both differential elements and the line between their centers. For this purpose, the emitting is assumed as isotropic.

8.1.2 View Factor Algebra

The view factor algebra is a combination of basic configuration factors between surfaces with different geometries and some fundamental relations between them (Howell 1982a):

- *Superposition rules*: Two superposition rules could be defined for the view factors to surfaces. They help to estimate the view factors which cannot be evaluated directly.

  Rule 1: The product of the view factor F_{i-j} from a surface i to surface j and the area A_i of surface i is equal to the sum of the products of the view factors from the parts of surface i to surface j and their areas.

$$F_{i-j}A_i = \sum_{k=1}^{N} F_{i_k - j}A_{i_k} \qquad (8.1\text{c})$$

  Rule 2: The view factor F_{i-j} from a surface i to surface j is equal to the sum of the view factors from the surface i to the parts of the surface j.

Fig. 8.1 Defining geometry for configuration factor (Howell 1982a)

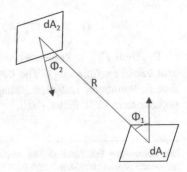

$$F_{i-j} = \sum_{k=1}^{N} F_{i-j_k} \tag{8.1d}$$

- *Summation rule*: The sum of the view factors from a given surface in an enclosure, including the possible self-view factor for concave surfaces, is 1.
- *Reciprocity relation*: A reciprocity relation between two opposite view factors of two isotropic emitting / receiving surfaces exists and allows the calculation of a view factor from the knowledge of its reciprocal:

$$A_i F_{i-j} = A_j F_{j-i} \tag{8.1e}$$

- *Bounding*: View factors are bounded to $0 \leq F_{i-j} \leq 1$ by definition.

New derivative view factors can be computed from a set of known factors with the help of the mentioned fundamental relations.

8.2 View Factor Between Two Surfaces with a Common Edge

This case has a number of applications, such as solar energy reflected from the ground and falling on a sloping roof, solar thermal collectors for water or air, or photovoltaic modules. The reflected irradiance I_R, received by an inclined surface from the uniform reflecting ground, is estimated with Eq. (8.2a).

$$I_R = \rho I_{GH} VF \tag{8.2a}$$

where ρ is the albedo value, VF is the view factor, and I_{GH} is the global horizontal irradiance.

8.2.1 View Factor from a Sloped Surface with Infinite Width to Infinite Uniform Horizon

8.2.1.1 View Factor from a Sloped Surface with Infinite Width to Infinite Uniform Horizon

Problem definition

Determine the view factor from an inclined plane A_1 to another infinite horizontal plane A_2 (Fig. 8.2). This task is often used to determine the amount of reflected solar radiation from a horizontal surface with isotropic reflectivity.

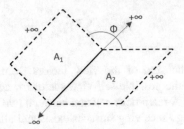

Fig. 8.2 Defining geometry for case with VF from inclined plane to infinite uniform horizon. The dashed lines symbolize infinity in the corresponding directions

Solution

The view factor from the sloping surface of infinite width to an infinite uniform horizon is to be estimated with the Eq. (8.2b). Input data is the included angle Φ.

$$VF = (1 + \cos \Phi)/2 \qquad (8.2b)$$

The task is solved in Module1 of the Ex08-01.xlsm. The Data sheet includes a column with different values of the included angle Φ and the corresponding results for VF. The module includes three functions and a main executable sub-routine:

- The function *Degrees_To_Radians*() converts an angle from degrees to radians.
- The function *VF_infinite_surfaces*() estimates the view factor from a sloped surface with infinite width to infinite uniform horizon.
- The procedure *VF_08_01_Main*() uses the above function for each angle ϕ in the Data sheet.

Discussion

Note that for $\Phi = 90°$, the VF is 0.5, for $\Phi = 0°$, the VF is 1, and for $\Phi = 180°$, the VF is 0. The first procedure *Degrees_To_Radians*() will be used again in the following examples.

8.2.1.2 View Factor from a Sloped Surface with Infinite Width to Finite Uniform Horizon

Problem definition

Estimate the view factor from an inclined plane A_1 with infinite to another finite horizontal surface A_2 (Fig. 8.3). This task is often used to determine the amount of reflected solar radiation from a horizontal surface with an isotropic reflectivity to a vertical or inclined surface of building's wall or PV.

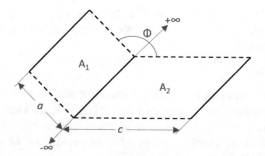

Fig. 8.3 Defining geometry for case with VF from inclined sloped surface to finite uniform horizon. The dashed lines symbolize infinity in the corresponding directions

Solution

The view factor from a sloping surface of infinite width to a finite uniform horizon is estimated with Eq. (8.2c), where $L = c/a$. Input data is a, c, and included angle Φ.

$$VF = \frac{1 + L - \sqrt{L^2 - 2L \cos \Phi + 1}}{2} \tag{8.2c}$$

The task is solved in Module1 of the Ex08-02.xlsm. The Data sheet includes three columns with input data: a, c and the included angle ϕ and an additional column with the corresponding results for VF. The module includes the already mentioned function *Degrees_To_Radians()* and other procedures as follows:

- The function *VF_sloped_surface_to_infinite_horizon()* estimates the view factor from sloped surface with infinite width to finite uniform horizon. It receives the value of the argument $L = c/a$ and the included angle Φ (variable *Fi* in the VBA code).
- The procedure *VF_08_02_Main()* uses the above function for each dataset of a, c, and angle Φ in the Data sheet.

Discussion

Note that for $\Phi = 180°$, the VF is always 0. For $\Phi = 0°$, the VF varies: for $a < c$, the VF is 1; and for $a > c$, the VF is c/a.

8.2.2 View Factor from a Sloped Surface with Finite Width to Infinite Uniform Horizon

This case has many applications, for example, in determining the heat exchange between building walls and the ground, between closely spaced walls of the same building, between the surfaces of different buildings, between solar panels and closely spaced surfaces, and so on.

8.2.2.1 View Factor from a Vertical Surface with Finite Width to Finite Uniform Horizon Estimated with an Analytical Approach

Problem definition

Estimate the view factor from a vertical rectangular surface A_1 with finite width to another finite horizontal rectangular surface A_2 (Fig. 8.4). This task is often used to determine the amount of reflected solar radiation from a horizontal surface with isotropic reflectivity to a vertical building façade, or window, or vertical PV.

Solution

The VF from a vertical rectangular surface to a finite uniform horizon is given as Eq. (8.2d). The input data is: a, b, c (see Fig. 8.4).

For any two perpendicular surfaces with a common edge b (Fig. 8.4), the VF is to be estimated with Eq. (8.2d), according (Howell 1982c), where $N = c/b$ and $L = a / b$:

$$F_{1-2} = \frac{1}{\pi L} \left(\begin{array}{l} L\tan^{-1}\left(\frac{1}{L}\right) + N\tan^{-1}\left(\frac{1}{N}\right) - \sqrt{N^2 + L^2}\tan^{-1}\left(\frac{1}{\sqrt{N^2+L^2}}\right) \\ + \frac{1}{4}\left\{ \ln\left[\frac{(1+L^2)(1+N^2)}{1+L^2+N^2}\right] + L^2\ln\left[\frac{L^2(1+N^2+L^2)}{(1+L^2)(1+N^2)}\right] + N^2\ln\left[\frac{N^2(1+N^2+L^2)}{(1+N^2)(N^2+L^2)}\right] \right\} \end{array} \right)$$

(8.2d)

The task is solved in Module1 of the Ex08-03.xlsm. The Data sheet includes three columns with input data: a, b, and an additional column with the corresponding results for VF. The module includes a function and an executable public procedure:

- The function *VF_90degrees*() estimates the view factor from the vertical rectangle to a horizontal rectangle with a common edge. It receives the values of both arguments $N = c/b$ and $L = a/b$.
- The procedure *VF_08_03_Main*() uses the above function for each dataset of a, b, and c in the Data sheet.

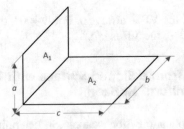

Fig. 8.4 Defining geometry for case with VF from a vertical surface to finite uniform horizon

Discussion

Note that for $a = c = 1$ and $b = 1{,}000{,}000$, the VF is 0.292783. We received the same value (0.292893) in the Ex08-02.xlsm (case #16 with $a = c = 1$ and $Fi = 90°$). For values of b, higher than 10^3, the horizon can be considered infinite; thus, the previous example works quite well and much faster for such cases.

8.2.2.2   View Factor from a Sloped Surface with Finite Width to Finite Uniform Horizon Estimated with an Analytical Approach

Problem definition

Estimate the view factor from a sloped rectangular surface A_1 with finite width to another finite horizontal rectangular surface A_2 (Fig. 8.5). This task is often used to determine the amount of reflected solar radiation from a horizontal surface with isotropic reflectivity to an inclined PV panel, with $\Phi > 90°$.

Solution

The VF from a sloped rectangular surface to a horizontal rectangle with a common edge is estimated with Eq. (8.2e), where $A = c/b$, $B = a/b$, $C = A^2 + B^2 - 2AB\cos\Phi$ and $D = \sqrt{1 + A^2 \sin^2\Phi}$ (Howell 1982d):

$$
\begin{aligned}
F_{1-2} = &-\frac{\sin 2\Phi}{4\pi B}\left[AB\sin\Phi + \left(\frac{\pi}{2} - \Phi\right)(A^2 + B^2) + B^2\tan^{-1}\left(\frac{A - B\cos\Phi}{B\sin\Phi}\right) + A^2\tan^{-1}\left(\frac{B - A\cos\Phi}{A\sin\Phi}\right)\right] \\
&+ \frac{\sin^2\Phi}{4\pi B}\left\{\left(\frac{2}{\sin^2\Phi} - 1\right)\ln\left[\frac{(1 + A^2)(1 + B^2)}{1 + C}\right] + B^2\ln\left[\frac{B^2(1 + C)}{C(1 + B^2)}\right] + A^2\ln\left[\frac{A^2(1 + A^2)^{\cos 2\Phi}}{C(1 + C)^{\cos 2\Phi}}\right]\right\} \\
&+ \frac{1}{\pi}\tan^{-1}\left(\frac{1}{B}\right) + \frac{A}{\pi B}\tan^{-1}\left(\frac{1}{A}\right) - \frac{\sqrt{C}}{\pi B}\tan^{-1}\left(\frac{1}{\sqrt{C}}\right) \\
&+ \frac{\sin\Phi\sin 2\Phi}{2\pi B}AD\left[\tan^{-1}\left(\frac{A\cos\Phi}{D}\right) + \tan^{-1}\left(\frac{B - A\cos\Phi}{D}\right)\right] \\
&+ \frac{\cos\Phi}{\pi B}\int_0^B \sqrt{1 + z^2\sin^2\Phi}\left[\tan^{-1}\left(\frac{z\cos\Phi}{\sqrt{1 + z^2\sin^2\Phi}}\right) + \tan^{-1}\left(\frac{A - z\cos\Phi}{\sqrt{1 + z^2\sin^2\Phi}}\right)\right]dz
\end{aligned}
\qquad (8.2e)
$$

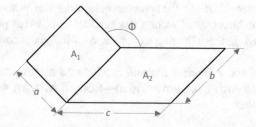

Fig. 8.5 Defining geometry for case with VF from a sloped rectangle to a horizontal rectangle that shares a common edge

The last part of Eq. (8.2e) is an unsolvable integral. This explains why a complete analytical solution does not exist. The view factor F_{1-2} can be estimated partially analytically and partially numerically. However, we will call this solution *analytical* because the analytical part is predominant.

The task is solved in Module1 of the Ex08-04.xlsm. The Data sheet includes four columns with input data: *a*, *b*, *c*, angle *Fi* (Φ), and an additional column with the corresponding results for VF. The module includes the already mentioned function *Degrees_To_Radians*() and other procedures, as follows:

- The function *VF_inclined_analytic*() estimates the view factor from a sloped rectangle to a horizontal rectangle with a common edge. It receives the values of arguments $A = c/b$, $B = a/b$, and the included angle *Fi* (Φ).
- The procedure *VF_08_04_Main*() uses the above function for each dataset of *a*, *b*, *c*, and *Fi* in the Data sheet.

Discussion

The previous sub-Sect. 8.2.2.1 presents a special sub-case of the present case, at an angle of 90°. Therefore, the results in the table in the Data sheet that correspond to a right angle between both surfaces ($\Phi = 90°$) should be identical to the results in the previous example.

Note the specific case of the VF from a vertical square to a horizontal adjusted square ($a = 1$, $b = 1$, $c = 1$ and $\Phi = 90°$). This is the same case as when we estimate the VF from one face of a cube to some other adjusted face. According to case #4 in the table, the resulting VF is 0.20004. This could help us to calculate the VF between two opposite cube's sides with the help of the summation rule of view factor algebra. It should be equal to $1 - 4 \times 0.200044$, i.e. 0.199825.

8.2.3   View Factor for Generalized Inclined-Rectangle Arrangement

Determining view factor in more complex configurations of reflecting and radiating surfaces is a very useful task that can be solved with the help of the view factor algebra. In (Muneer et al. 2015), the approach of how this can be done for simpler and more complex combinations of rectangular surfaces in different planes forming an angle is considered. In the following examples, we will look at some of these cases.

8.2.3.1   View Factor from a Sloped Surface to a Horizontal Surface with the Same Width—Solved with an Analytical Approach

Problem definition

Estimate the view factor from an inclined rectangular surface with finite width to another horizontal rectangular surface (Fig. 8.6). Both surfaces do not share a

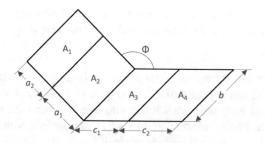

Fig. 8.6 Defining geometry for surfaces with the same width

common edge but have the same width. This task is often used to determine the amount of reflected solar radiation from a horizontal surface with isotropic reflectivity to an inclined or vertical surface (parts of façade, window, PV). Input data is: a_1, a_2, b, c_1, c_2, and included angle Φ. The values a_1 and c_1 are the distances between the common edge of both planes to the corresponding closer parallel edges of the receiving and emitting receiving rectangles. Estimate the VF_{1-3}, VF_{2-4}, and VF_{1-4} (see again Fig. 8.6).

Solution

The solution uses the rules of view factor algebra and the Eq. (8.2e), applied in the function *VF_inclined_analytic()*.

The task is solved in Module1 of the Ex08-05.xlsm. The Data sheet includes inputs $a1$, $a2$, $c1$, $c2$, b, and angle *Fi* (in degrees). The module includes the already used functions *Degrees_To_Radians()* and *VF_inclined_analytic()*, and one more executable main procedure:

- The procedure *VF_08_05_Main()* calculates for each dataset in the Data sheet the following view factors: VF_{12-34}, VF_{12-3}, VF_{12-4}, VF_{2-34}, VF_{2-3}, VF_{2-4} and VF_{1_4}.

Discussion

The calculating part of the procedure *VF_08_05_Main()* is relatively short. There is no need for additional functions. The value of VF_{12-34} is estimated with calling of the function *VF_inclined_analytic(a1 + a2, c1 + c2, b, Fi_deg)*, and the value of VF_{12-3} is estimated with the function *VF_inclined_analytic(a1 + a2, c1, b, Fi_deg)*. These values are used to calculate VF_{12-4} as a difference between VF_{12-34} and VF_{12-3}, using the superposition rule 2 of VFA. Then, the value of VF_{2-34} is estimated with the function *VF_inclined_analytic(a1, c1 + c2, b, Fi_deg)*, and the value of VF_{2-3} is estimated with the function *VF_inclined_analytic(a1, c1, b, Fi_deg)*. The VF_{2-4} is calculated as a difference between VF_{2-34} and VF_{2-3}, using the superposition rule 2 of VFA again. Finally, the VF_{1-4} is calculated using the superposition rule 1.

8.2.3.2 View Factor for Generalized Inclined-Rectangle Arrangement, Estimated with an Analytic Approach

Problem definition

Estimate the view factor from an inclined rectangular surface A_1 with finite width to another horizontal rectangular surface $A_{3'}$ (Fig. 8.7). Both surfaces do not share a common edge and have different widths. This task is often used to determine the amount of reflected solar radiation from a horizontal surface with isotropic reflectivity to an inclined or vertical surface (parts of façade, window, PV). The analytic approach should be used. The input data is the coordinates a_{1L}, a_{1U}, $c_{3'L}$, $c_{3'U}$, b_{1L}, b_{1U}, $d_{3'L}$, $d_{3'U}$, and included angle Φ. The coordinates a_{1L}, a_{1U} are along the x_1 axis, the coordinates $c_{3'L}$, $c_{3'U}$ are along the x_2 axis, and the coordinates b_{1L}, b_{1U}, $d_{3'L}$, $d_{3'U}$ are along the $y_1 = y_2$ axes.

Solution

The solution uses the rules of view factor algebra and the Eq. (8.2e), applied in the function *VF_inclined_analytic*(). Let us have two rectangular surfaces with a common edge, separated by given angle Φ, and let each of them have six rectangular parts: $A_{123456} = A_1 + A_2 + A_3 + A_4 + A_5 + A_6$ and $A_{1'2'3'4'5'6'} = A_{1'} + A_{2'} + A_{3'} + A_{4'} + A_{5'} + A_{6'}$ (Fig. 8.7). The solving Eq. (8.2f) is presented in (Holman 1992) for two perpendicular surfaces and in (Muneer et al. 2015) for an inclined-rectangle arrangement.

$$
A_1 F_{1-3'} = \frac{1}{2} \left(
\begin{array}{l}
K_{(123456)^2} - K_{(1256)^2} - K_{(2345)^2} + K_{(25)^2} - K_{(4,5,6)-(1'2'3'4'5'6')} + K_{(56)-(1'2'5'6')} \\
+ K_{(45)-(2'3'4'5')} - K_{5-(2'5')} - K_{(123456)-(4'5'6')} + K_{(1256)-(5'6')} + K_{(2345)-(4'5')} \\
- K_{(25)-5'} + K_{(4,5,6)^2} - K_{(56)^2} - K_{(45)^2} + K_{5^2}
\end{array}
\right)
$$

$$(8.2f)$$

where the K terms are defined by $K_{m-n} = A_m F_{m-n}$ and $K_{(m)^2} = A_m F_{m-m'}$.

The task is solved in Module1 of the Ex08-06.xlsm. The Data sheet includes inputs a_{1L}, a_{1U}, $c_{3'L}$, $c_{3'U}$, b_{1L}, b_{1U}, $d_{3'L}$, $d_{3'U}$, and included angle Φ (*Fi*) in degrees.

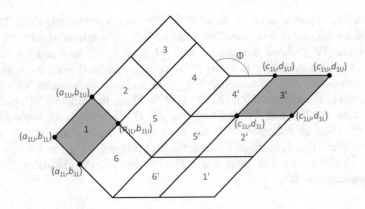

Fig. 8.7 Defining geometry for generalized inclined-rectangle arrangement

The module includes the already used functions *Degrees_To_Radians()* and *VF_inclined_analytic()* and two new procedures:

- The function *VF_13()* estimates the VF between the rectangles A_1 and $A_{3'}$ (see Fig. 8.7), using Eq. (8.2f).
- The procedure *VF_08_06_Main()* calculates for each dataset in the Data sheet the needed VF.

Discussion

Using the coordinates of the emitting and receiving rectangle and Eq. (8.2f), we can solve the problem of any mutual arrangement of the two rectangles, including those specified in the previous example, which can be used to verify the current VBA code. In cases #1 and #2 of the present Ex08-06.xlsm, we calculate VF_{1-4} and VF_{2-4} from case #1 of the previous Ex08-05.xlsm. The equal results prove the correctness of this analytic approach.

8.2.3.3   View Factor for a Generalized Inclined-Rectangle Arrangement, with Finite Element Approach with a Uniform Grid

Problem definition

Estimate the view factor from an inclined rectangular surface A_1 with finite width to another horizontal rectangular surface $A_{3'}$ (Fig. 8.8). Both surfaces do not share a common edge and have different widths. The finite element approach with a uniform grid should be used. We consider the rectangular surfaces A_1 and $A_{3'}$ as composed of many very small rectangular cells (Fig. 8.8), and we could use numeric integration to estimate the VF between both surfaces with a small loss of accuracy.

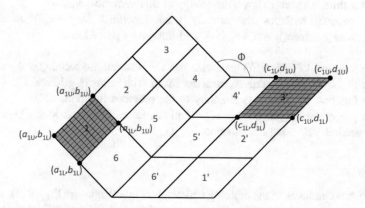

Fig. 8.8 Defining geometry for generalized inclined-rectangle arrangement with finite element approach with uniform grid

The input data is the coordinates a_{1L}, a_{1U}, $c_{3'L}$, $c_{3'U}$, b_{1L}, b_{1U}, $d_{3'L}$, $d_{3'U}$, and included angle Φ, like the inputs in the Ex08-06.xlsm file. The finite elements approach needs four more inputs—cell's sizes in both directions of both surfaces: *delx1*, *dely1*, *delx2*, and *dely2*.

Solution

The solution uses the finite element approach and Eq. (8.2g) from (Muneer et al. 2015).

$$F_{i-j} = \frac{\sin^2 \Phi}{\pi.Na.Nb} \sum_{i_1=1}^{Na} \sum_{i_2=1}^{Nb} \sum_{j_1=1}^{Nc} \sum_{j_2=1}^{Nd} \frac{x_i x_j}{\left[x_i^2 + x_j^2 - 2x_i x_j \cos \Phi + (y_i - y_j)^2\right]^2} \Delta c \Delta d$$

$$(8.2g)$$

where $\Delta a = a/Na$, $\Delta b = b/Nb$, $\Delta c = c/Nc$, $\Delta d = d/Nd$ and Na, Nb, Nc, Nd are the numbers of intervals for the numeric integration in each dimension. The coordinates of each cell's center are: for surface $i - x_i = (i_1-0.5)\Delta a$; $y_i = (i_2-0.5)\Delta b$; for surface $j - x_j = (j_1-0.5)\Delta c$; $y_j = (j_2-0.5)\Delta d$. This solution has a significant advantage—it can be easily adapted for any disposition of both rectangular surfaces. In the same time, it has two serious disadvantages—it gives an approximate result and needs a lot of computing time to avoid this with large numbers of intervals. The approach could be adapted for nonuniform emitting or reflecting of the second surface.

Each rectangle is divided into many small rectangular cells with equal sizes in both dimensions—*delx1* × *dely1* for the receiving surface A_1 and *delx2* × *dely2* for the emitting surface $A_{3'}$.

The task is solved in Module1 of the Ex08-07.xlsm. The Data sheet includes inputs a_{1L}, a_{1U}, $c_{3'L}$, $c_{3'U}$, b_{1L}, b_{1U}, $d_{3'L}$, $d_{3'U}$, and included angle Φ (*Fi*) in degrees, and the cells' sizes *delx1*, *dely1*, *delx1*, and *dely1*. There are two columns with corresponding results for VF, estimated with UG approach and analytically, a two other columns—with error and number of iterations. The number of iterations is linearly related to used computer time. The error shows the quality of this numerical solution.

The module includes the already used functions *Degrees_To_Radians()*, *VF_inclined_analytic()*, and *VF_13()*, and three new procedures:

- The function *VF_13_UG()* estimates the VF between the rectangles A_1 and $A_{3'}$ (see Fig. 8.7), using finite element approach with uniform grid and Eq. (8.2f).
- The function *VF_funct()* is ancillary to the previous function.
- The procedure *VF_08_07_Main()* calculates for each dataset in the Data sheet the needed VF, using the function *VF_13_UG()*.

Discussion

Using the coordinates of the emitting and receiving rectangle and Eq. (8.2f), we can solve the problem of any mutual arrangement of the two rectangles, including those specified in the previous example, which can be used to verify the current VBA code. In cases #1 and #2 of the present Ex08-07.xlsm, we calculate VF_{1-4} and VF_{2-4} from case #1 of the Ex08-05.xlsm. The last column with the estimated error shows high accuracy for the distant rectangles, with an error, under 1%.

8.2.3.4 View Factor for Generalized Inclined-Rectangle Arrangement, with Finite Element Approach with Uniform Grid, Combined with Monte Carlo Method

Problem definition

Estimate the view factor from an inclined rectangular surface A_1 with finite width to another horizontal rectangular surface $A_{3'}$ (Fig. 8.8), using a finite element with uniform grid approach, combined with Monte Carlo method. This method takes its name from the famous casinos located in the city of Monte Carlo. It uses repeated random sampling to make numerical estimations of unknown parameters and enables modeling of complex systems with random variables. There are many applications of Monte Carlo methods.

In the present context, the user may select a Computation Reduction Factor (CRF) to save computational time (Muneer et al. 2020). For example, if a reduction factor of 5 is chosen, then the elemental view factor will be evaluated only if the randomly generated number falls between 0 and 1/5 (the normal range is between 0 and 1). The final view factor will be the average of all the individual cell-based view factors.

The input data is the coordinates a_{1L}, a_{1U}, $c_{3'L}$, $c_{3'U}$, b_{1L}, b_{1U}, $d_{3'L}$, $d_{3'U}$, and included angle Φ, like the inputs in Ex08-06.xlsm file. The finite elements approach needs four more inputs: *delx1*, *dely1*, *delx2*, and *dely2*. The last column of the inputs is CRF—a positive integer that varies from 1 to a higher number. There are two columns with corresponding results for VF, estimated with UGMC approach and analytically, two other columns—with error and number of iterations. As in the previous example, the number of iterations is linearly related to used computer time. The error shows the quality of this numerical solution.

Solution

The task is solved in Module1 of the Ex08-08.xlsm. The Data sheet includes inputs a_{1L}, a_{1U}, $c_{3'L}$, $c_{3'U}$, b_{1L}, b_{1U}, $d_{3'L}$, $d_{3'U}$, and included angle Φ *(Fi)* in degrees, and the cells' sizes *delx1*, *dely1*, *delx1*, and *dely1*, and a Computation Reduction Factor (CRF). The module includes the already used functions *Degrees_To_Radians()*, *VF_inclined_analytic()*, *VF_13()* and *VF_funct()*, and two new procedure:

- The function *VF_13_UGMC()* estimates the VF between the rectangles A_1 and $A_{3'}$ (see Fig. 8.8), using finite element approach with uniform grid, combined with Monte Carlo method.
- The main executable procedure *VF_08_08_Main()* calculates for each dataset in the Data sheet the needed VF, using the function *VF_13_UGMC()*.

Discussion

In this solution, as in the previous one, the main executable procedure calculates an error for each dataset by comparing the result of the *VF_13UGMC()* procedure with the most accurate analytical result. It turns out that even if we reduce the number of iterations for the calculation by 10 times, the accuracy is still good enough, and the error is below 1%. The error is especially low for arrangements where the rectangles are relatively distant from each other. Thus, thanks to the Monte Carlo method, we obtain sufficiently accurate results in less computational time.

8.3 View Factor Between Two Parallel Surfaces

The determination of VF between two parallel surfaces has great application in determining the energy exchange between opposite walls of an urban street canyon. It is a place where the street is flanked by buildings on both sides creating a canyon-like environment. This is the case with most streets in city centers.

The view factors discussed below can be used to determine both the emitted and reflected solar radiation by two parallel surfaces. The reflected irradiance I_R, received by a vertical surface from the uniform reflecting opposite vertical surfaces, is estimated again with the Eq. (8.2a), mentioned in Sect. 8.2.

8.3.1 View Factor Between Two Parallel Surfaces with Infinite Width and Finite Height

8.3.1.1 View Factor Between Two Parallel Surfaces with Infinite Width and Finite Height, Directly Opposite

Problem definition

Determine the view factor from a vertical surface A_1 to another parallel surface A_2 (Fig. 8.9). Both surfaces are with infinite width and finite height and are directly opposite. This task is often used to determine the amount of reflected solar radiation between two parallel walls of an urban canyon in the case of isotropic reflectivity.

Solution

The view factor between two parallel surfaces with infinite width and finite height, directly opposite, is to be estimated with Eq. (8.3a), for $H = c\,/\,a$:

$$VF_{1-2} = VF_{2-1} = \sqrt{1 + H^2} - H \qquad (8.3a)$$

The task is solved in Module1 of the Ex08-09.xlsm. The Data sheet includes a column with different values of a and c, and the corresponding estimated results for VF. The module includes one function and a main executable sub-routine:

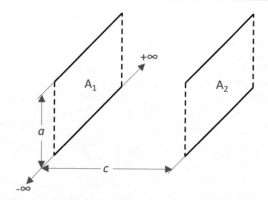

Fig. 8.9 Defining geometry for case with VF between two parallel surfaces with infinite width and finite height, directly opposite. The dashed lines symbolize infinity in the corresponding directions

- The function *VF_infinite_parallel_surfaces_directly_opposite*() estimates the view factor between both parallel surfaces.
- The procedure *VF_08_09_Main*() uses the above function for each dataset in the Data sheet.

Discussion

VF is applicable when calculating the reflected irradiance received by a vertical wall from the opposite infinite vertical wall with uniform reflectivity. It is calculated in the same way, regardless of whether the reflecting surface is parallel or at an angle to the receiving surface. Therefore, Eq. (8.2a) is also applicable for parallel surfaces, and there VF is calculated with the procedure *VF_infinite_parallel_surfaces_directly_opposite*() or with the procedures discussed in the following examples.

8.3.1.2   View Factor Between Two Parallel Surfaces with Infinite Width and Different Finite Height

Problem definition

Determine the view factor from a vertical surface A_1 to another parallel surface A_2 (Fig. 8.10). Both surfaces are with infinite width and finite height and are directly opposite. This task is often used to determine the amount of reflected solar radiation between two parallel walls of an urban canyon in the case of isotropic reflectivity. The inputs are the heights a_1 and a_2, distance c between both surfaces, and the vertical distance a_3 between the bottom edges of the surfaces.

Solution

The view factor between two parallel surfaces with infinite width and finite height, not directly opposite, is to be estimated with Eq. (8.3b):

Fig. 8.10 Defining geometry for case with VF between two parallel surfaces with infinite width and finite height, not directly opposite. The dashed lines symbolize infinity in the corresponding directions

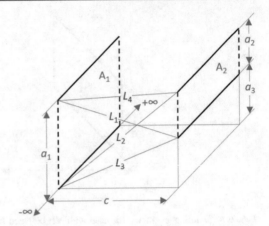

$$F_{1-2} = \frac{L_1 + L_2 - L_3 - L_4}{2a_1} \tag{8.3b}$$

where $L_1 = \sqrt{c^2 + (a_1 - a_3)^2}$, $L_2 = \sqrt{c^2 + (a_2 + a_3)^2}$, $L_3 = \sqrt{c^2 + a_3^2}$ and $L_4 = \sqrt{c^2 + (a_2 + a_3 - a_1)^2}$.

The task is solved in Module1 of the Ex08-10.xlsm. The Data sheet includes a column with different values of a_1, a_2, a_3, and c, and the corresponding estimated results for VF. The module includes one function and a main executable sub-routine:

- The function *VF_infinite_parallel_surfaces()* estimates the view factor between both parallel surfaces.
- The procedure *VF_08_10_Main()* uses the above function for each dataset in the Data sheet.

Discussion

The previous example is a special case of the present example where $a_1 = a_2$ and $a_3 = 0$. Two possible applications of this example are estimating the VF between two parallel infinitely long rows of PV panels or between two infinitely long walls of an urban canyon.

8.3.2 View Factor Between Two Directly Opposite Parallel Rectangles with Equal Size

Problem definition

Determine the view factor from a vertical rectangle A_1 to another parallel rectangle A_2 (Fig. 8.11). Both rectangles have equal finite widths and finite heights and are

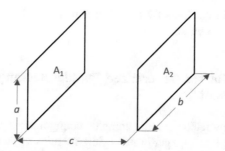

Fig. 8.11 Defining geometry for case with VF between two parallel, directly opposite surfaces with dimensions a and b, and distance c between them

directly opposite. This task is often used to determine the amount of reflected solar radiation between two finite parallel walls of an urban canyon in the case of isotropic reflectivity. The inputs are both dimensions of the rectangles—a and b, and the distance c between them.

Solution

The view factor between two parallel surfaces with infinite width and finite height, not directly opposite, is to be estimated with Eq. (8.3c)—(Howell 1982b):

$$F_{1-2} = F_{2-1} = \frac{2}{\pi XY} \left(\begin{array}{l} X\sqrt{1+Y^2}\tan^{-1}\left(\frac{X}{\sqrt{1+Y^2}}\right) + Y\sqrt{1+X^2}\tan^{-1}\left(\frac{Y}{\sqrt{1+X^2}}\right) \\ -X\tan^{-1}(X) - Y\tan^{-1}(Y) + \ln\left[\frac{(1+X^2)(1+Y^2)}{1+X^2+Y^2}\right]^{1/2} \end{array} \right)$$

$$(8.3c)$$

where $X = a/c$ and $Y = b/c$.

The task is solved in Module1 of the Ex08-11.xlsm. The Data sheet includes four columns with different values of a, b, and c, and the corresponding estimated results for VF. The module includes one function and a main executable sub-routine:

- The function *VF_opposite_parallel_rectangles()* estimates the view factor between both directly opposite parallel rectangles.
- The executable procedure *VF_08_11_Main()* uses the above function for each dataset in the Data sheet.

Discussion

The calculated values of VF in the file Ex08-11.xlsm show that for the same values of a and b, but with increasing values of the distance c between the two surfaces, VF between them gradually decreases from maximum 1 to minimum 0, and the relationship is not linear. At very large values of b, the calculation results are very close to the results in the Ex08-09.xlsm file; that is, the effect is like VF between two surfaces of infinite width.

8.3.3 View Factor for Generalized Parallel Rectangles Arrangement

Determining the view factor in more complex configurations of parallel reflecting and radiating surfaces can be performed using the rules of view factor algebra. In (Ivanova et al. 2016), simpler and more complex arrangements of parallel rectangular surfaces are considered. In the following examples, we will look at some of these cases.

8.3.3.1 View Factor Between Two Parallel Rectangles in General Arrangement

Problem definition

Determine the view factor from a vertical rectangle $A_{3'}$ to another parallel rectangle A_7 (Fig. 8.12). This task is often used to determine the amount of reflected solar radiation between two finite parallel walls of an urban canyon in the case of isotropic reflectivity. The inputs are the coordinates a3L, a3U, b3L, b3U, c7L, c7U, d7L, d7U, e1, and e2.

Solution

The solution uses the rules of view factor algebra and Eq. (8.3c), applied in the function *VF_opposite_parallel_rectangles*(). Let us have two parallel rectangular surfaces, and let each of them have nine rectangular parts, where $A_1 = A_{1'}$, $A_2 = A_{2'}$ etc. (Fig. 8.12). The solution is in Eq. (8.3d) presented in (Holman 1992).

Fig. 8.12 Defining geometry for case with VF between two parallel rectangles A_3 and $A_{7'}$ in general arrangement

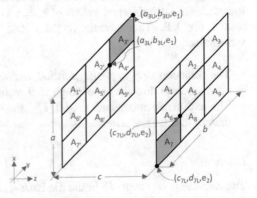

$$F_{3'-7} = \frac{1}{A_{3'}} \left(\begin{array}{l} K_{(1,2,3,4,5,6,7,8,9)^2} - K_{(2,3,4,5,8,9)^2} - K_{(1,2,5,6,7,8)^2} + K_{(2,5,8)^2} - K_{(4,5,6,7,8,9)^2} \\ + K_{(4,5,8,9)^2} + K_{(5,6,7,8)^2} - K_{(5,8)^2} - K_{(1,2,3,4,5,6)^2} + K_{(2,3,4,5)^2} + K_{(1,2,5,6)^2} \\ - K_{(2,5)^2} + K_{(4,5,6)^2} - K_{(4,5)^2} - K_{(5,6)^2} + K_{(5)^2} \end{array} \right) / 2$$

$$(8.3d)$$

where the K terms are defined by $K_{m-n} = A_m F_{m-n}$ and $K_{(m)^2} = A_m F_{m-m'}$. The term $K_{(m)^2}$ corresponds to basic VF, estimated with Eq. (8.3c).

The task is solved in Module1 of the Ex08-12.xlsm. The Data sheet includes inputs $a_{3'L}$, $a_{3'U}$, $b_{3'L}$, $b_{3'U}$, c_{7L}, c_{7U}, d_{7L}, d_{7U}, e_1, and e_2. The module includes the already mentioned function *VF_opposite_parallel_rectangles()* and two new procedures:

- The function *VF_37()* estimates the view factor between two parallel rectangles in general arrangement (VF from rectangle $A_{3'}$ to rectangle A_7, see Fig. 8.12).
- The executable procedure *VF_08_12_Main()* uses the above function for each dataset in the Data sheet.

Discussion

Using the coordinates of the emitting and receiving rectangle and Eq. (8.3d), we can solve the problem of any mutual arrangement of the two parallel rectangles with collinear edges, including those specified in the previous example which can be used to verify the current VBA code. The cases #1 to #4 in the present Ex08-12.xlsm correspond to examples in (Suryanarayana 1995); the equal results prove the correctness of the approach.

8.3.3.2   View Factor Between Two Parallel Rectangles in General Arrangement, with Finite Element Approach with a Uniform Grid

Problem definition

Determine the view factor from a vertical rectangle $A_{3'}$ to another parallel rectangle A_7 (Fig. 8.13). The inputs are the coordinates a3L, a3U, b3L, b3U, c7L, c7U, d7L, d7U, e1, and e2. The finite elements approach needs four more inputs: *delx1, dely1, delx2,* and *dely2,* like in Ex-08-08.xlsm file.

Solution

The task is solved in Module1 of the Ex08-13.xlsm. The Data sheet includes inputs $a_{3'L}$, $a_{3'U}$, $b_{3'L}$, $b_{3'U}$, c_{7L}, c_{7U}, d_{7L}, d_{7U}, e_1, and e_2, and cells' dimensions: *delx1, dely1, delx2,* and *dely2.* The module includes the already mentioned functions *VF_opposite_parallel_rectangles()* and *VF_37()*, and two new procedures:

- The function *VF_37_UG()* estimates the view factor between two parallel rectangles in general arrangement with finite element approach, using all cells of a uniform grid.

Fig. 8.13 Defining geometry for case with VF between two parallel rectangles $A_{3'}$ and A_7 in general arrangement, with finite element approach with a uniform grid

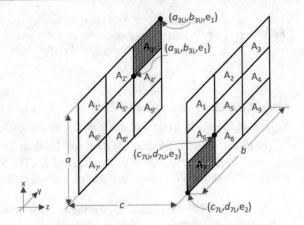

- The executable procedure *VF_08_13_Main()* uses the above functions for each dataset in the Data sheet.

Discussion

The cases #1 to #4 in the present Ex08-13.xlsm correspond to examples in (Suryanarayana 1995). Compared to the analytic result, estimated in Ex08-12.xlsm, the error is under 0.02%. It is observed that when calculating the VF between parallel surfaces with finite element approach with uniform grid, the error is smaller than when the same approach is applied to surfaces that are inclined to each other and have a common edge.

8.3.3.3 View Factor Between Two Parallel Rectangles in General Arrangement, with Finite Element Approach with Uniform Grid and Monte Carlo Method

Problem definition

Determine the view factor from a vertical rectangle $A_{3'}$ to another parallel rectangle A_7 (see Fig. 8.13). The inputs are the coordinates a3L, a3U, b3L, b3U, c7L, c7U, d7L, d7U, e1, e2, *delx1, dely1, delx2* and *dely2*, and Calculation Reduction Factor (CRF).

Solution

The task is solved in Module1 of the Ex08-14.xlsm. The Data sheet includes inputs $a_{3'L}$, $a_{3'U}$, $b_{3'L}$, $b_{3'U}$, c_{7L}, c_{7U}, d_{7L}, d_{7U}, e_1, and e_2, and cells' dimensions: *delx1, dely1, delx2* and *dely2*, and CRF. The module includes the already mentioned functions *VF_opposite_parallel_rectangles()* and *VF_37()*, and two new procedures:

- The function *VF_37_UGMC()* estimates the view factor between two parallel rectangles in general arrangement with finite element approach, using a uniform grid, combined with Monte Carlo method.

- The executable procedure *VF_08_14_Main*() uses the above functions for each dataset in the Data sheet.

Discussion

In this solution, as in the previous one, the main executable procedure calculates an error for each dataset by comparing the result of the *VF_37UGMC*() procedure with the most accurate analytical result. Here, too, the same feature is observed as when applying this method to two surfaces at an angle in the Ex08-08.xlsm file. Even if we reduce the number of iterations for the calculation by 10 times and more, the accuracy is still high enough, and the error is below 1%. Thus, thanks to the Monte Carlo method, we obtain sufficiently accurate results in less computational time.

8.4 Conclusion

The task of determining view factors has many applications in heat transfer and in building physics. The value of VF is needed for calculating the energy exchange between the walls and the base of an urban street canyon or for calculating the reflected solar radiation from the terrain to the walls of a building or to solar panels. It can be applied to determine the amount of reflected light. In this chapter, we have considered analytical and numerical ways to determine VF between surfaces inclined relative to each other, with or without a common edge, as well as between parallel surfaces in different arrangements. The surfaces could be finite or infinite in one or two directions. All examples are implemented using appropriate VBA procedures in Excel files.

All figures in this chapter are created within MS Excel using VBA procedures, described in Chap. 4.

References

J.P. Holman, *Heat Transfer*, 7th edn. (McGraw-Hill, New York, 1992)

J.R. Howell, *A Catalog of Radiation Heat Transfer—Configuration Factors. Introduction.* McGraw-Hill (1982a). Available on: http://www.thermalradiation.net/intro.html

J.R. Howell, *A Catalog of Radiation Heat Transfer—Configuration Factors. C-11: Identical, Parallel, Directly Opposed Rectangles* (1982b). Available on: http://www.thermalradiation.net/sectionc/C-11.html

J.R. Howell, *A Catalog of Radiation Heat Transfer—Configuration Factors. C-14: Two Finite Rectangles of Same Length, Having One Common Edge, and at an Angle of 90° to Each Other* (1982c). McGraw-Hill. Available on: http://www.thermalradiation.net/sectionc/C-14.html

J.R. Howell, *A Catalog of Radiation Heat Transfer—Configuration Factors. C-16: Two Rectangles with One Common Edge and Included Angle of Φ* (1982d). McGraw-Hill. Available on: http://www.thermalradiation.net/sectionc/C-16.html

S. Ivanova, T. Muneer, Finite-element heat-transfer computations for parallel surfaces with uniform or non-uniform emitting. J. Renew. Sustain. Energy **8**(1), 015102-1–015102-16 (2016)

T. Muneer, S. Ivanova, Y. Kotak, M. Gul, Finite-element view-factor computations for radiant energy exchanges. J. Renew. Sustain. Energy, 033108-1–033108-20 (2015)

T. Muneer, S. Ivanova, Efficient routines for obtaining radiation view-factor for non-uniform horizons. Energies **13**,(2551), 1–17.https://doi.org/10.3390/en13102551

N.V. Suryanarayana, *Solution Manual for Engineering Heat Transfer* (West Publishing Company, New York, 1995)

Artificial Neural Networks in Excel

9

9.1 Introduction to Artificial Neural Networks

The mystery of human intelligence has long excited people. In the twentieth century, a more specific approach based on the '*neural network*' concept began to be applied to this problem. A *neural network* (natural or artificial) is a collection of many connected neurons. A neuron is a simple thing and does nothing most of the time. However, if it receives a signal through its incoming channels (dendrites), it excites and transmits slightly modified to the neurons to which it is connected (axons) through its synapses. The neuron itself has no special capabilities. However, connected to many other neurons, it forms a network capable of recognizing and analyzing images, events, memories, and taking actions. This happens to the simplest living beings that have a brain bud, as well as to animals, birds, and higher primates, including humans. In combination with many years of training, the huge number of neurons leads to higher nervous activity.

In order to study and model human thinking, psychologist Rosenblatt (1962) proposed a concept for an electronic device called a *perceptron*. This is the first development that marks the beginning of the so-called *artificial neural networks* (ANNs), which consist of a number of interconnected elements called *neurons*. Like natural networks, neurons are organized into *layers* that can be 2, 3, or more. The simplest network consists of (i) an *input layer*, which includes the input data that is fed to the neural network; (ii) a *hidden layer*, which allows to model complex relationships between input and output; and (iii) an *output layer*, where we obtain the results of the network operation (Fig. 9.1). The more layers there are in the neural network, the more complex tasks it can perform. The individual neurons in layers interact with each other through connections, called *weights*, which determine the strength and influence of the connected neurons.

Supplementary Information The online version contains supplementary material available at https://doi.org/10.1007/978-3-030-94085-0_9.

Fig. 9.1 Exemplary neural
network architecture with one
hidden layer

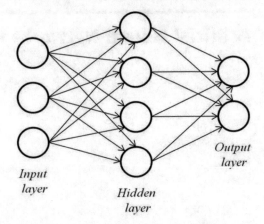

*Input
layer*

*Hidden
layer*

*Output
layer*

However, it is astonishing how effective even the simplest neural network with
only three layers can be, such as the one in Fig. 9.1, if it is trained, i.e. configured to
work with input from a problem area. Even a relatively limited number of repetitions of input data processing lead the neural network to relatively efficient work.

Neural networks have one, two, or more inner *hidden layers* that can be connected in different ways. It has been proved mathematically by Csáji (2001) with
the help of the *Universal Approximation Theorem* that ANN with more than one
hidden layer with a sufficient number of neurons can model the behavior of any
existing function. A network with more than one hidden layer is trained through
so-called *deep learning*.

The state (value) of each neuron v in the first hidden layer is defined as the sum
of the products of each input x_j with the corresponding weighting factor w_j and the
deviation b (*bias*)—see Eq. (9.1a):

$$v = \sum_{j=1}^{m} w_j x_j + b = \sum_{j=0}^{m} w_j x_j \tag{9.1a}$$

In addition to the weight coefficients, neural networks are determined by the
activation functions of every layer, without the input (Csáji, 2001). These functions
can be very simple or more complex, for example, *linear*—Eq. (9.1b); *log-sigmoid*
—Eq. (9.1c); *hyperbolic tangent*—Eq. (9.1d); *ReLu*—Eq. (9.1e); and others.

$$f(x) = cx \tag{9.1b}$$

$$f(x) = 1/(1 + e^{-x}) \tag{9.1c}$$

$$f(x) = 2/(1 + e^{-2x}) - 1 = 2sigmoid(2x) - 1 \tag{9.1d}$$

$$f(x) = \max(0, x) \tag{9.1e}$$

9.1.1   Principle of ANN Training with Back Propagation

To become useful, the ANN must be trained. Unlike other computer systems, in ANN, this is done through examples, most often accumulated through observations and experiments rather than by programming rules of operation. The training is to adjust the weights so that they lead to the desired results. There are different training methods. The most common is the *back propagation* (Rumelhart et al. 1986), short for '*backward propagation of errors*', in which the difference between the desired and the obtained value is calculated for each output neuron. This creates an error signal, which is fed back to the input layer and, on its way, changes the weights so that the next time you activate the ANN, the error will decrease even more. This is repeated many times. The one-time training with the all dataset is called an *epoch*. Usually, ANN is trained with several thousand (up to millions) repetitions (epochs).

Once the ANN is trained, it must be 'fed' with new data that was not used during its training. This is a type of testing of the ANN to determine if it is able to perform the desired job.

An important point in the development of ANN is to determine the number of hidden layers and their neurons. There are no fixed rules or technology; most often, the number is determined by experimentation, looking for the smallest error.

Here is a summary of the technology for training ANN by error propagation. First, the variables that participate as input parameters for the ANN must be defined. It is desirable that the number of input parameters is not very large. However, it should be noted that if the input parameters include those that do not affect the desired results, after training, they will correspond to zero or almost zero weights values; that is, the neural network during its training distinguishes the influencing from the non-influencing input parameters (Goh 1995).

This is followed by data collection for training and testing of ANN. This requires the accumulation of datasets of input parameters and the corresponding result (or results). The training must have a sufficiently large and representative database. It is recommended for training to have 5–10 records for each weighting factor (Hammerstrom 1993). Because neural networks are trained in linear interconnections more effectively, one of the goals of data generation is to reduce nonlinearity. If a given data X is reciprocal to the result, it is more efficient to use $(1/X)$ (Goh 1995).

There is one more reason for the need for the pre-processing of the data before using it for training in the ANN. This is the use of a log-sigmoid (or other similar) activation function. It leads to values of the output of each neuron in the range from 0 to 1. This requires normalization of the target values before the training. If a variable V varies between the minimum and maximum values, respectively V_{min} and V_{max}, the normalized value A is determined by Eq. (9.1f):

$$A = (V - V_{min})/(V_{max} - V_{min}) \qquad (9.1f)$$

The input datasets and results thus prepared are divided into two groups: (1) for training the ANN and (2) for testing the trained ANN, in a ratio of 2 to 1. Training begins with random initial values of neuron weights. The ANN then feeds on the training data and gradually adjusts its coefficients using the back propagation. Training continues until the mean square error for all training datasets in an epoch falls below a certain, predetermined minimum value.

There are no hard rules for determining the number of neurons in the hidden layer. If it is too small, the ANN may not be able to solve the problem and find the correct match between the input data and the results. Too many neurons increase training time and sometimes lead to worse results. The training usually begins with a small number of neurons, which can be increased, if the error remains high.

After the training, the ANN is tested with a pre-separated part of the data, with the aim of keeping the mean square error below a preset value.

After training and testing, the neural network can be run by feeding new series of data, on the basis of which it calculates the required results. It is very important that the input data submitted is in the same range as the input data during the training of the ANN.

The weight coefficients can be positive (corresponding to excitatory connections between neurons) and negative (suppressive connections).

9.1.2   Features of the Use of ANN. Areas of Application

In other modeling approaches, it is necessary to know the mathematical relationship between input data and results. This is not necessary for ANN. If there is input that is irrelevant to the result, in its training, the ANN assigns them zero or close to zero weighting coefficients. These input variables can then be removed from the model (Goh 1995). Another feature is that quite often, the input data is accompanied by 'noise', and the ANN is trained to work with it. When new, better, noise-free data emerges, the ANN can be retrained. A useful feature of ANN is that once the weights w_j and the bias b are determined, they can be used in any environment or integrated into small programs.

The biggest drawback of ANN is the inability to trace and explain step by step the logic of how to get from the input data to the results. Another disadvantage is that their training requires multiple repetitions.

A third disadvantage is that it happens that when searching for a solution to reach a random local minimum of the error function. In this case, efforts and purposeful intervention are needed to continue the search, and it is not certain that there is another, even smaller minimum.

The main tasks, which are solved by the methods of artificial intelligence, in particular by ANN, are (**Wikipedia: Artificial neural network**):

- Functional/regression analysis, including predictive models, approximation, and modeling;
- Classification, pattern and sequence recognition, decision making;
- Data processing, including clustering, filtering, allocation, and compression;
- Speech recognition and generation, image recognition;

- Traffic management and routing;
- Design automation.

9.2 Neural Network with One Hidden Layer in VBA for Regression Analysis

There are many software products for working with ANN. One of the most popular is MATLAB's Neural Network Toolbox (Zhao et al. 2010). It has a graphical interface and allows the user to create his activation functions, and is programmable in the environment of MATLAB.

However, in this chapter, we want to demonstrate that the modeling of neurons and a neural network with a hidden layer is not a so difficult task. Thus, we will show the implementation of an example of ANN with back propagation. This type of network performs very well in the field of regression analysis (**Wikipedia: Regression analysis**)—a field in mathematical statistics in which the possible functional dependencies between two or more random variables are studied and evaluated. The aim is to determine whether there is a relationship between these random variables and if so, to find a function that describes it accurately enough.

Regression analysis does not answer the question of what are the reasons for the relationship between the quantities but only shows the mutual relations between them, i.e. to establish how one quantity Y depends on the values of several others—$x_1, x_2 \dots x_n$—Eq. (9.2a)

$$Y = f(x_1, x_2, ..., x_n) \tag{9.2a}$$

In regression analysis, we are not interested in the essence of the modeled process and the specific meaning of the data, but we aim to establish the relationship between the input data and the required value, trying to cover the definition area of both input data and of the required value.

For illustrative purposes, we will further develop and gradually improve a sample program for the training and testing of a neural network.

As an example, we will train a neural network to calculate the view factor VF_{1-2} from surface A_1 to surface A_2, different approaches to which we discussed in Chap. 8. Let the two rectangular surfaces A_1 and A_2 have a common edge and an included angle Φ between them, which varies from $0°$ to $180°$ (Fig. 9.2). The data needed to calculate this VF_{1-2} is:

a—transverse size of the rectangle A_1,

c—transverse size of the rectangle A_2,

b—longitudinal size of both rectangles, i.e. size of the common edge between them,

Φ—angle between both considered surfaces.

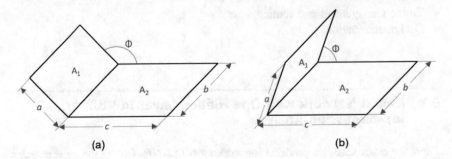

Fig. 9.2 Two plane surfaces with a common edge and included angle Φ: **a** $\Phi > 90°$, **b** $\Phi < 90°$

Instead of the variables a, b, and c, Eq. (8.2e) in Chap. 8 works with two new variables $A = c/b$ and $B = a/b$. This practically means that we could cover all variants of the mutual relations of c/b and a/b, assuming size $b = 1$. This reduces the number of required input data and their combinations.

In such tasks, for the purposes of regression analysis, very often, there is used the log-sigmoid function—Eq. (9.1c). It returns a normalized result for the value Y (from 0 to 1). In this case, when calculating the VF, the result is also generally between 0 and 1, so we do not need to normalize it before training. Normalization is not mandatory for the input values, but sometimes it helps to reach the desired learning result faster, i.e. to find out what the relationship is between the input variables x_1, x_2,... x_n, and the result Y.

To perform the task, we must perform the following steps:

- Preparation of the complete database for the training of the neural network;
- Dividing the data into two parts—for training and testing of the network;
- Creating a neural network training program;
- Neural network testing;
- Using the trained ANN.

9.2.1 Preparation of the Complete Neural Network Training Database

What could be the origin of the training data in regression analysis? One possible source of this data is from continuous measurements and observations. The second possible source is from modeling with a proven model, usually too complex to be widely used. In both cases, it is desirable that the data covers as much of the definition area as possible, both of the input data and of the result.

The data for the training of the neural network must be prepared in the form of a series of records that contain information both about the input data and the final result for the VF. In our example, this means that multiple sets of values of a, c, Φ, and the corresponding calculated values of VF must be prepared.

Problem definition

Create a program that generates the complete database (records) at values of:

- size a—to be changed from 1/5 to 5 in steps of 0.2 (25 values);
- size c—to be changed from 1/5 to 5 in steps of 0.2 (25 values);
- angle Φ—to be changed from 0 to 180 with step 5 (37 values).

In the general case, a and c should be able to change from 0 to $+\infty$, but because the neural network is trained with examples, this would increase the size of the training database too much, so we limit the maximum values of a and c to 5 times the size of the common edge b between the two surfaces. Note that for size $a = 0$, the first surface is a straight line and, the view factor cannot be defined.

In this situation, the number of records in the complete ANN training database must be $25 \times 25 \times 37 = 23{,}125$. Use the function *VF_inclined_analytic*() from Ex08-04.xlsm in Chap. 8 to calculate the required VF.

Solution and Discussion

In the Ex09-01.xlsm file, in Module1, the task is solved in 2 ways—with two seemingly similar procedures:

The *CreateDatabase_bad_solution*() procedure uses real control variables a and c to control the loops to implement all combinations. This is a practice that can lead to problems, and therefore, instead of the expected 23,125 records in the saved file VF-incl-bad.txt, there are only 21,312; that is, the task is not solved correctly. The records with $a = 5$ or $c = 5$ are missing.

The procedure *CreateDatabase_recommended*() uses integer loop control variables and calculates the values of a and c from them. Finally, in the saved file VF-incl-5x5.txt are recorded all the necessary records, 23,125 in number.

9.2.2 Dividing the Complete Neural Network Training Database into Two Parts

Problem definition

The created records in the complete database should be divided at random—about 2/3 of them will be used for training, about 1/3—for testing the neural network after training.

Create a procedure that splits the data as instructed and creates two new files, one for training and one for testing.

Solution and Discussion

The solution is shown in Module1 of Ex09-02.xlsm file. The *Split_Data*() procedure reads the VF-incl-5x5.txt file from the Files-for-Chapter-9 folder and randomly divides the records in a 2 to 1 ratio by writing them to the same folder in two files VF-incl-5x5-training.txt and VF-incl-5x5-testing.txt.

9.2.3 Neural Network Training

Problem definition

Create the necessary procedures for training a neural network with three layers—input, hidden, and output with one result. All network training parameters are to be set in the Data sheet. Read the training and testing data from the relevant files created in sub-section 9.2.2.

Solution

The solution is written in Module1 of the file Ex09-03.xlsm and contains a neural network with three layers—input, hidden, and output layer with one result.

The purpose of training in each step is to reduce the error, i.e. the difference between the state of the ANN output and the required final target value, corresponding to each input record. This is done through the method *back propagation*, which in turn adjusts all connections between neurons in order to reduce the error. In this example, both activation functions are log-sigmoid—see Eq. (9.1c).

The following Fig. 9.3 shows the table for ANN training parameters:

- Training data—location and name of the training data file;
- Testing data—location and name of the testing data file;
- Inputs—number of input data in one record;
- Hiddens—number of neurons in the hidden layer of ANN;
- Learning_rate—learning speed, with an initial value of 0.5;
- enhance—a variable that controls the learning process, with a value of 0.01;

	A	B	C
1	Training data	Files-for-Chapter-9\VF-incl-5x5-training.txt	Train ANN
2	Testing data	Files-for-Chapter-9\VF-incl-5x5-testing.txt	
3	inputs	3	Test ANN
4	hiddens	10	
5	Learning rate	0.5	
6	enhance	0.01	
7	Number of epochs	2000	
8	Training RMSE	9.72%	
9	Testing RMSE	8.70%	

Fig. 9.3 Contents of the ANN training parameter table in the file Ex09-03.xlsm

Table 9.1   Procedures *Train*() and *Run*() in Module1 of file Ex09-03.xlsm

```
Private Function Train(InputValue () As Double, Target As Double) As Double
Dim Deltas_ih () As Double
Dim net_ih () As Double
ReDim Deltas_ih (1 To hiddens)
ReDim net_ih (1 To hiddens)
Dim Deltas_ho As Double
Dim net_ho As Double
Dim i As Integer , j As Integer, Sum As Double

    For i = 1 To hiddens
       Sum = 0
       For j = 1 To inputs
          Sum = Sum + InputValue (j) * Weights_ih (j, i)
       Next j
       Sum = Sum + Bias_ih (i)
       net_ih (i) = 1 / (1 + Exp (-Sum))
    Next i

'Calculate the net for the final output layer.
    Sum = 0
    For i = 1 To hiddens
       Sum = Sum + net_ih (i) * Weights_ho (i)
    Next
    Sum = Sum + Bias_ho
    net_ho = 1 / (1 + Exp (-Sum))

'Backpropagation
    Deltas_ho = (net_ho + enhance) * (1 - net_ho + enhance) * (net_ho - Target)
    Bias_ho = Bias_ho - Learning_Rate * Deltas_ho

    For i = 1 To hiddens
       Deltas_ih (i) = Deltas_ho * Weights_ho (i) * (net_ih (i) + enhance) * (1 - net_ih (i) + enhance)
       Weights_ho (i) = Weights_ho (i) - Learning_Rate * Deltas_ho * net_ih (i)
       For j = 1 To inputs
          Weights_ih (j, i) = Weights_ih (j, i) – Learning_Rate * Deltas_ih (i) * InputValue (j)
       Next
       Bias_ih (i) = Bias_ih (i) - Learning_Rate * Deltas_ih (i)
    Next
    Train = net_ho
End Function

Private Function Run (InputValue () As Double) As Double
Dim net_ih () As Double
ReDim net_ih (1 To hiddens)

Dim Sum #, i%, j%
    For i = 1 To hiddens
       Sum = 0
       For j = 1 To inputs
          Sum = Sum + InputValue (j) * Weights_ih (j, i)
       Next
       Sum = Sum + Bias_ih (i)
       net_ih (i) = 1 / (1 + Exp (-Sum))
```

(continued)

Table 9.1 (continued)

```
    Next
    Sum = 0
    For i = 1 To hiddens
      Sum = Sum + net_ih (i) * Weights_ho (i)
    Next
    Sum = Sum + Bias_ho
    Run = 1 / (1 + Exp (-Sum))
End Function

Function Normalize_Value (Value As Double, MinV As Double, MaxV As Double) As Double
  Normalize_Value = (Value - MinV) / (MaxV - MinV)
End Function
```

- Number of epochs in which ANN training is repeated;
- Training RMSE—ANN error determined during training;
- Testing RMSE—ANN error detected during testing.

The first simplified version of the neural network training program in Module1 includes the description of the following functions and procedures (see Table 9.1):

- Procedure *Reset_Weights*() pre-declares all necessary arrays and initializes all parameters and weights of the neural network.
- Procedure *Display_Weights*() displays the values of all parameters (weights, deviations, etc.) of the neural network in the Weights sheet.
- Function *Train*() is carried out actual training ANN method back propagation. It receives as arguments the input data as a one-dimensional array *InputValue*() and the desired result *Target*.
- Function *Run*() estimates the value of the result for submitted input data using already preset parameters of the trained neural network.
- Function *Normalize_Value*() returns the normalized value (between 0 and 1) of an argument, using the minimum and maximum value of its definition area, using the Eq. (9.1f).
- Procedure *Load_Data*() reads the training database file.
- Function *Calc_RMSE*() calculates the RMSE (Root Mean Square Error) for the entire dataset of the values calculated with ANN compared to the expected results.
- Global executable procedure *Train_ANN_Main*() calls all the above-mentioned procedures by reading the learning parameters from the Data sheet, initializing the neural network parameters, loading the training database from a file, normalizing the values of Φ, a, and c, and training the neural network in the specified number of epochs. Finally, it calculates and displays the value of RMSE after training.

- Procedure *Read_Weights*() reads the values of the weights and other parameters of the neural network from the Weights sheet. The procedure is called by the following procedure *Test_ANN_Main*().
- Global executable procedure *Test_ANN_Main*() initializes the parameters of the neural network, reads the parameters of ANN from the Weights sheet, loads the database for testing from a file, normalizes the values of Φ, *a,* and *c*, calculates the result for each data record, and finally, displays the value of RMSE after testing.

Discussion

The procedure *Reset_Weights*() initializes all weighting factors and biases of ANN as random numbers between 0 and 1 using the VBA function *Rnd*(). It is not desirable for the initial values of the weighting factors to be all fixed as 0 or 0.5, because this makes learning difficult.

The *Train*() function receives as arguments the values of each entry in the training database. It has three parts. In the first part, the state *net_ih*(i) of each neuron *i* is calculated using the input data in the array *InputValue*() and their respective *Weights*(j, i), as well as the deviation *Bias_ih*(i) according to Eq. (9.1a). The value obtained is fed to the log-sigmoid function to activate the neuron. In the second part of the function, the value of the result is calculated using the state of each neuron *net_ih*(i) and the weighting factor *Weights_ho*(i) of the relationship between the neuron and the result as well as the deviation *Bias_ho*. The obtained value is passed to a second activation function (in this case, again log-sigmoid) for activating the output neuron and calculating its final state *net_ho*.

In the third part, the back propagation of the error (*net_ho – Target*) is realized, which is propagated in reverse order of the calculation in the first two parts on all interneuronal connections. The variable *Learning_Rate* controls the learning speed, i.e. how big are the steps to minimize the error. Usual values at the beginning of the training of this variable are from 0.5 to 0.1. A second variable that also controls this process is *enhance*, most often with a value of 0.01.

The following function *Run*() gets the input data values for each record in the database as arguments. It starts as the function *Train*() with the same first two parts. The difference between the two functions is that *Run*() lacks back propagation.

For the two public executable procedures, *Train_ANN_Main*() and *Test_A-NN_Main*() in the Data sheet, two call buttons are made (Fig. 9.3). Pressing the '*Train ANN*' button starts the training procedure for a given number of epochs (in the example—2000). As a result, the neural network is trained by the data, in the order in which they are set in the database file, i.e. in the order of their generation with cycles, described in sub-section 9.2.1.

RMSE (Root Mean Square Error) is calculated by Eq. (9.2b), where Y_T is the target result, and Y_C is the calculated result.

$$RMSE = \sqrt{\frac{\sum (Y_T - Y_C)^2}{\sum Y_T}} \qquad (9.2b)$$

The resulting value of RMSE with training data after 2000 epochs is 9.72% (this is an approximate value that may vary due to the element of randomness in the initial values of the neural network parameters). Pressing the 'Test ANN' button starts the Test_ANN_Main() procedure, which reads the test data file, which is about half the number of training records. The resultant value of RMSE is 8.70% (sample value, it may be different).

9.2.4 Improving the Algorithm for Training of ANN

As the number of learning epochs increases, the accuracy of the trained neural network can be expected to improve. The problem is that the improvement is not linear but attenuating (Fig. 9.6). However, there are many approaches developed to improve ANN training. One of them is especially simple and effective, and therefore, we will use it to improve the program developed by us for training neural networks in the previous example.

ANN can 'remember' and 'forget'. If the data is sorted by the values of Φ, a, and c parameters, with the advancement of training within an epoch, the network begins to forget what was learned with the previous values. Therefore, a neural network to which data is fed in a shuffled order is trained faster and more efficiently.

Problem definition

Develop a new, improved version of the program that, after reading the training data, shuffles it before submitting it to the ANN for training. Add the ability to work with various activation functions. Track and record training and testing time, as well as RSME training and testing errors. Write in a separate sheet and file the parameters of the trained ANN.

Solution

An improved version of the previous example is saved in Module1 of the Ex09-04.xlsm file. In addition to some of the procedures used in the previous Ex09-03.xlsm, new procedures and functions have been added and others have been slightly changed.

Figure 9.4 shows the table with parameters for the training of ANN—extended compared to the previous example, below are listed only the added new parameters:

- Weights—location and the name of the file, where the parameters of the trained ANN will be saved;
- TypeActivation1 and TypeActivation2—type of the first and second activation functions in the *Train()* function. The possible values are: 1—for *logistic sigmoid function*, 2—for *linear function*; 3—for *logarithmic sigmoid function*; 4—for *hyperbolic tangent function*.

▲	A	B	C
1	Training data	Files-for-Chapter-9\VF-incl-5x5-training.txt	Train ANN
2	Testing data	Files-for-Chapter-9\VF-incl-5x5-testing.txt	
3	Weights	Files-for-Chapter-9\VF-incl-5x5-weights.txt	Test ANN
4	inputs	3	
5	hiddens	10	Draw Comparison Figure
6	Learning rate	0.5	
7	enhance	0.001	
8	TypeActivation1	1	
9	TypeActivation2	1	
10	Number of epochs	2000	
11	Total epochs	4000	
12	Randomize	Yes	
13	Start time	15:16:15	
14	End Time	15:19:17	
15	Time for training	0:03:02	
16	Training RMSE	2.09%	
17	Testing RMSE	1.99%	

Fig. 9.4 Contents of the ANN training parameters table in the Ex09-04.xlsm file, with record shuffling enabled. Time for training in 2000 epochs—3:02 min, for processor Intel(R) CPU @ 2.90 GHz

- Total epochs—number of completed epochs for training so far;
- Randomize—Yes/No—indicates whether the database should be shuffled or arranged;
- Start time, End time—beginning and end of the training period;
- Time for training.

The procedures *Reset_Weights()*, *Normalize_Value()*, *Load_Data()*, and *Calc_RMSE()* remain unchanged. The new procedures are:

- The *Save_Weights()* procedure saves the weights and other parameters of the ANN in the Files-for-Chapter-9 folder as a file named VF-incl-weights.txt, whose name is specified in the Data sheet.
- The *Activation()* function expects the activation type as an integer from 1 to 4. It calculates the value of the function according to Eqs. (9.1.2) to (9.1.5). This function is called by the main two functions of ANN – *Train()* and *Run()*.
- The function *UnNormalize_Value()* unnormalizes a variable, returns it to its normal range, and is the opposite of the function *Normalize_Value()*.
- The *Randomize_Data()* procedure shuffles the data read by the file, generating two random numbers (from 1 to the number of records) and swapping the records that match them. This is done as many times as there are records in the database.
- The public executable procedure *Draw_Comparison()* compares the accuracy of the values of target VF with those calculated by the neural network VF.

The following procedures have been changed:

- The *Display_Weights*() procedure displays the weights and other parameters of the ANN in a different place than in the previous example.
- The *Read_Weights*() procedure reads the weights and other parameters of the ANN from their new location.
- The *Train*() function is similar to the function of the same name in Ex09-03.xlsm but calls the *Activation*() function, which allows the use of four different types of the activation function.
- The *Run*() function is similar to the function in Ex09-03.xlsm, but is extended by calling the *Activation*() function.
- Public executable procedure *Train_ANN_Main*() reads all training parameters from the Data sheet, including the two types of activation function, whether the data should be shuffled, the name of the resulting weights' file. In addition to the tasks performed by the analogous procedure in Ex09-03.xlsm, the start and end times of the training are detected, and the RMSE at the end of the training and the total number of training periods accumulated so far are displayed.
- The public executable procedure *Test_ANN_Main*() is analogous to the procedure of the same name in Ex09-03.xlsm but displays the RMSE from testing at a new position.

Discussion

The Data sheet contains three buttons to start the three executable public procedures. Pressing the 'Train ANN' button starts the training procedure for a given number of epochs (again 2000), but this time the data is shuffled. The resulting value of RMSE with training data is 2.10% (i.e. an example value that may vary due to the element of randomness in the initial values of the neural network parameters and the random mixing of the data). Pressing the 'Test ANN' button starts the *Test_ANN_Main*() procedure, which reads the test data file, which is about half the number of training records, normalizes the data in the three columns, and tests the ANN. The resultant value of RMSE is 2.04% (sample value, it may be different).

Again, it is important to point out that with the different random initial values of all weighting factors and the random mixing of the data after reading them, an element of randomness is introduced in the result. The search for a minimum error value is rather a search for a local error minimum. Pressing the 'Train ANN' button once starts the procedure *Train_ANN_Main*(), which results in training that ends with a set of network parameters. Restarting it continues the training from where it reached at the end of the previous training. It should lead to a reduction in the RMSE error, but in reality, this does not always happen. Repeating or increasing the number of epochs several times may or may not lead to a smaller error. To restart the training from the beginning, it is enough to reset the *Total_Epochs* variable in the Data sheet.

Another improvement that was added to the procedure *Train_ANN_Main*() in this example is controlling the learning speed by the *Learning_rate* variable. Its initial value is set to 0.5 in the Data sheet, but it decreases by 5% each time when the error value at the end of an epoch is greater than that of the previous epoch, indicating that ANN begins to move away from the local minimum of the error. Gradually reducing *Learning_rate* allows the ANN to approach the local minimum without skipping it and without starting to move away from it.

The sheet Weights of the Ex09-04.xlsm file (Fig. 9.5) shows how to work with the weights and other parameters of the neural network determined as a result of the training (shown in green). The first green column contains the values of 10 weighting factors for the first variable (angle Φ, which has normalized values from 0 to 1)—one for each neuron; the second column contains the coefficients for the normalized values of a; and the third—for the normalized values of c. Look at the formulas in the orange cells. The first column calculates the sum of the products of the weights by the input data from the record under consideration. In the second column, the value of the activation function (in this case, log-sigmoid) is calculated with an argument to the value to the left of each cell in the column. The third column calculates the product of the elements in the second column by the corresponding element in the *Weights_out* column. These values are then added vertically, and the *bias* value written at the bottom of the first green column is added to them. The normalized result is obtained by applying the second activation function (again log-sigmoid) to *Sum + Bias*. In case of minimum and maximum of the desired result, other than 0 and 1, unnormalization is required, the formula for

	A	B	C	D	E	F	G	H	I	J
1		Fi	a	c	Target VF	How the neural network works...				
2	Original Inputs	90	1	1	0.20004					
3	Normalized Inputs	0.5	0.2	0.2	0.20004					
4		Weights & Biasses						Calculations		
5		Weight_Fi	Weight_a	Weight_c	Weight_out	Bias		Sum W*inputs	Activation	Act*Weight_out
6	Neuron 1	-3.1477	-1.2790	-15.3089	-13.3409	-0.1367		-5.0281	0.0065	-0.0868
7	Neuron 2	-3.5341	-10.7784	-8.3985	10.6248	-0.0191		-5.6215	0.0036	0.0383
8	Neuron 3	-1.9576	-6.1420	-0.3900	5.2125	-0.3517		-2.6369	0.0668	0.3482
9	Neuron 4	-5.1022	-0.4862	0.0161	12.1355	5.7174		3.0723	0.9557	11.5983
10	Neuron 5	74.3718	-0.3473	1.4543	-31.9813	1.0183		38.4256	1.0000	-31.9813
11	Neuron 6	25.8170	-0.6980	2.1409	10.4300	-0.6249		12.5722	1.0000	10.4299
12	Neuron 7	-5.0608	3.0043	-7.5475	-1.2863	1.1495		-2.2896	0.0920	-0.1183
13	Neuron 8	-2.1040	-2.8541	0.8377	1.9025	0.8328		-0.6225	0.3492	0.6644
14	Neuron 9	-6.8715	0.0193	-0.4202	3.5181	0.8564		-2.6595	0.0654	0.2301
15	Neuron 10	34.3257	-0.1919	1.6827	-4.0238	-2.6413		14.8332	1.0000	-4.0238
16	Bias	11.5114							Sum+Bias:	-1.3896
17									Normalized Result:	0.1995
18								Unnormalized Final Result:		0.1995
19										
20		Target_minimum:	0						Error:	-0.29%
21		Target_maximum:	1							

Fig. 9.5 A table with the parameters of the trained ANN after 4000 epochs and an example of how it is used to obtain a result in the Ex09-04.xlsm file. The target VF is 0.20004, the ANN result is 0.1995, the error for these input values is −0.29%

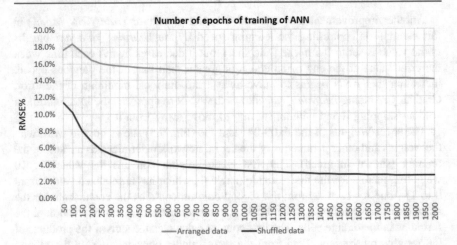

Fig. 9.6 Reducing the RMSE error as the number of ANN training epochs increases in the Training sheet of the Ex09-04.xlsm file

which is written in the dark orange cell. In this case, the normalized and unnormalized values coincide because VF varies from 0 to 1 (i.e. there is a minimum of 0 and a maximum of 1).

All this shows that once trained, the neural network (i.e. its coefficients and parameters) can be used very easily, quickly, and efficiently. The main operations that are performed in its work are addition, multiplication, and exponent. The time required to apply a neural network to a data record is directly proportional to the number of neurons in it. Therefore, there is a desire to reduce the number of neurons, but this increases the training time of the neural network. Therefore, an appropriate optimal number of neurons is sought.

To better understand the learning process, the program is expanded to display RMSE every 50 epochs when training online. This is illustrated in the graph of Fig. 9.6. The RMSE error is marked in the vertical direction, and its value in the horizontal direction decreases as the number of epochs increases. The graph in light gray is for the ANN training with the ordered data as in Ex09-03.xlsm and in dark gray for the shuffled data in Ex09-4.xlsm. It can be seen that in the second case, ANN is much more effective and completes the training in 2000 epochs with about 4–5 times less error. The other thing that can be seen in the graph of the ordered data is that the decrease in the error is not smooth with the progress of the number of epochs, but there may be an increase in the error at some points. This can happen not only with ordered data (light gray curve) but also with shuffled data (dark gray curve) in case of too high a value of *Learning_rate*. In such cases, the value of this variable begins to decrease (in the case of the light gray curve for 2000 epochs, *Learning_rate* decreases from 0.5 at the beginning to 0.0006 at the end). On the

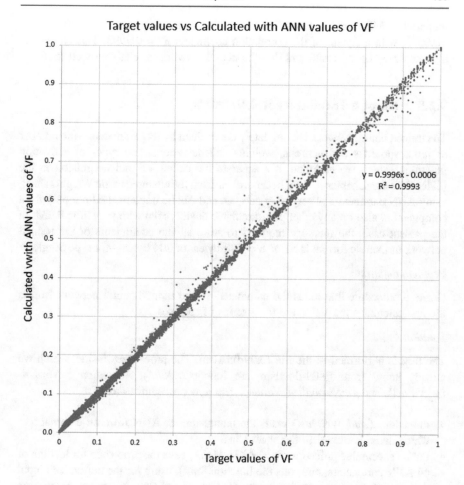

Target values vs Calculated with ANN values of VF

y = 0.9996x - 0.0006
R² = 0.9993

Fig. 9.7 Graph of the comparability of desired target values and calculated with ANN results after 4000 epochs, in the Comparison sheet of the file Ex09-04.xlsm

other hand, too low a value of this variable from the beginning leads to a lower learning speed, and the error curve decreases more slowly, which requires more learning epochs.

The third button in the Data sheet is the 'Draw Comparison Figure', which generates a list of testing input data, target VF, and calculated VF via ANN in the Comparison sheet. This data is used to create the image in Fig. 9.7, which shows the correspondence between the required VF values calculated by Eq. (8.2e) in Chap. 8 and those obtained during neural network testing. Each pixel corresponds to one record of test data. Ideally, the pixel should lie exactly on the diagonal line. We see that this is true with small deviations for most points. Since the graph was prepared in MS Excel, a trend line was added, whose equation is $y = 0.9996 \times -0.0006$, with a square of the correlation coefficient $R^2 = 0.9993$. This shows the high accuracy of the result and that the neural network is relatively well trained.

9.2.5   Use of a Trained Neural Network

The trained neural network is completely determined by its parameters—the number of input variables and neurons, weights and deviations, and types of activation functions, which we recorded in a separate file called VF-incl-weights.txt in the folder Files-for-Chapter-9. As it became clear from the example in the Weights sheet, using ANN parameters is very easy and fast and can be implemented not only for a computer but also for any other programmable device, even with very little RAM. In the present case, the memory required to store all the coefficients of the trained network in Double format is $51 \times 8 = 408$ bytes, or 204 bytes—for type Single.

Problem definition

Create a procedure that loads the parameters of the trained neural network from a file and calculates the VF value for a series of surfaces.

Solution

The task is performed in the file Ex09-05.xlsm. The procedures inside, which we already know from Ex09-04.xlsm, are functions *Run()*, *Activation()*, *Normalize_Value()*, and *UnNormalize_Value()*. The new procedures are:

- Procedure *Load_Weights()* reads the parameters of ANN from an external file, which name is placed in the sheet Data.
- Public executable procedure *Run_ANN_Main()* calls the procedure for loading of all ANN parameters, and calls the function *Run()*, using for the normalized input data in the same sheet: dimensions *a, b,* and *c* of two rectangles A_1 and A_2 (Fig. 9.8), and angle Φ between them.

▲	A	B	C	D	E	F	G	H	I
1	**Weights**	Files-for-Chapter-9\VF-incl-5x5-weights.txt							Run ANN
2	#	a	b	c	Fi	Target	ANN result	Error	
3	1	1	1	1	90	0.200044	0.199472	-0.29%	
4	2	50	50	60	90	0.211163	0.209834	-0.63%	
5	3	50	30	60	90	0.170278	0.167312	-1.74%	
6	4	50	20	60	90	0.138096	0.135883	-1.60%	
7	5	1	1	1	30	0.619028	0.595972	-3.72%	
8	6	1	1	1	45	0.483348	0.475614	-1.60%	
9	7	1	1	1	60	0.370905	0.363363	-2.03%	
10	8	1	1	1	90	0.200044	0.199472	-0.29%	
11	9	1	1	1	120	0.086615	0.087494	1.01%	
12	10	1	1	1	135	0.048309	0.047731	-1.20%	
13	11	1	1	1	150	0.021345	0.021572	1.06%	

Fig. 9.8 A table in Data sheet of the file Ex09-05.xlsm with the parameters *a, b, c,* and *Fi* (angle Φ), see Fig. 9.2, analytic calculated target values of VF (Muneer et al. 2015), calculated results with ANN, and errors

9.3   Conclusion

In this chapter, we have shown that modeling and using an artificial neural network with back propagation is not a difficult task. Step by step, we created a complete training database and divided it into two parts—for training and for testing the already trained neural network; then, we trained in 2000 epochs a simpler modeled neural network, with which we received RMSE 8.7% in testing. In the next better modeled neural network, we added a series of improvements: shuffling training data, managing the *Learning_rate* parameter to reduce learning speed when needed, adding the ability to change activation functions, and more. With such an improved network of 4000 epochs, we have achieved better training and RMSE below 2%. In a separate sheet, we showed an example how the calculated values of ANN weighting coefficients are used. The parameters of the trained network saved in a separate file were used in the last example for calculation of VF at different input data.

This chapter shows how quickly we can find the right relationship between input and results if we have experimental data or results from a more complex model. In just a few minutes and several thousand epochs, we can train a neural network to solve this problem with an accuracy of 1–2%. The interpreter-driven VBA programming language may not be best suited for more complex tasks—with larger datasets and more epochs because it is slower, but it makes it easier to track how the training of a neural network works and understand that the training of an artificial neural network is not an unbearably complex task. We hope that these examples will contribute to more people learn and use neural networks.

There are two main issues in solving similar problems with ANN. The first is that we cannot avoid the error in the solution, and the second is that a significant reduction in error requires many more epochs of training, and it is not possible to know in advance whether there is a solution with a smaller error. The approach is suitable for complex tasks that depend on many parameters that cannot be analyzed analytically, and at the same time, if there is accumulated experimental data to power the neural network during its training.

References

B.C. Csáji, Approximation with Artificial Neural Networks. MSc Thesis, Faculty of Sciences; EötvösLoránd University, Hungary, 2001

A.T.C. Goh, Back-propagation neural networks for modeling complex systems. Artif. Intell. Eng. **9**, 143–151 (1995)

D. Hammerstrom, Working with neural networks, in *IEEE Spectrum*, July (1993), pp. 46–53

T. Muneer, S. Ivanova, Y. Kotak, M. Gul, Finite-element view-factor computations for radiant energy exchanges. J. Renew. Sustain. Energy **7**, 033108 (2015)

F. Rosenblatt, *Principles of Neurodynamics: Perceptrons and the Theory of Brain Mechanisms* (Spartan Books, Washington, DC, 1962)

D.E. Rumelhart, G.E. Hinton, R.J. Williams, Learning representations by back-propagating errors. Nature **323**(6088), 533–536 (1986). Bibcode:1986 Natur.323..533R. https://doi.org/10.1038/323533a0.

Wikipedia: Artificial neural network. https://en.wikipedia.org/wiki/Artificial_neural_network

Wikipedia: Regression analysis. https://en.wikipedia.org/wiki/Regression_analysis

Z. Zhao, H. Xin, Y. Ren, X. Guo, Application and comparison of BP neural network algorithm in MATLAB, in *2010 International Conference on Measuring Technology and Mechatronics Automation* (2010), pp. 590–593. https://doi.org/10.1109/ICMTMA.2010.492

Fluid Mechanics

10

A fluid is defined as that substance that continually deforms under shear stress. Fluid mechanics is that branch of engineering science that involves in its understanding the application of laws of mechanics and thermodynamics along with the knowledge of fluid properties.

In this chapter, we shall use numerous examples to explore computations related to fluid friction through single and multiple pipe systems and pumping power required to motivate the flow.

Example 10.1 Obtaining pipe diameter for a given flow rate and pressure drop.

Problem definition

Tel-Jala Corporation of Faudiland has engaged Engineer Ahmed Kash to design a pipeline facility to transport oil across the 200 km desert. The desired flow rate of oil through the galvanized iron pipe is 10,800 m^3/h. The maximum pressure the pipe can handle is 84 bar and at no point in the pipeline should the pressure drop below 4 bar. Layout the procedure that Ahmed Kash has to undertake to find an appropriate pipe diameter.

Solution

Refer to Example 3.4 where an approximate and a precise procedure to compute fluid-flow friction factor, f was presented. The approximate method by Swamee and Jain is 99% accurate and hence that method shall be used in the present example (see Eq. 10.1.1). Note that 'f' is dependent upon Reynolds number which in turn depends on flow velocity and relative

---

Supplementary Information The online version contains supplementary material available at https://doi.org/10.1007/978-3-030-94085-0_10.

roughness which in turn depend on pipe diameter being provided. Hence the solution can only be obtained iteratively by assuming pipe diameter at the start of the computation chain which will have the steps shown in Textbox 10.1.

$$f = \frac{1.325}{\left[\ln\left(\frac{\varepsilon/D}{3.7}\right) + \frac{5.74}{Re^{0.9}}\right]^2} \quad (10.1.1)$$

Textbox 10.1 Computation chain to iteratively obtain pipe diameter.

1. Given data: Q, Δp, L, μ, ρ, ε
2. Assume pipe diameter, D
3. Calculate flow velocity, $V = 4Q/(\pi D^2)$
4. Calculate Reynolds number, $Re = \rho \cdot V \cdot D / \mu$
5. Calculate relative roughness = ε/D
6. Calculate friction factor f (Eq. 10.1.1)
7. Calculate pressure drop, = $\rho f L V^2/(2D)$
8. Check whether calculated pressure drop matches the value given in step 1. If not, repeat the entire procedure using an altered value of diameter, D

Launch the file Ex10-01.xlsm by double-clicking the file icon. The given data has been keyed-in rows 1–6 and rows 8–14 provide the computational chain. Cell B14 provides the value of error in estimating pressure drop for the presently chosen value of pipe diameter, i.e. 1 m. The object of this exercise is to select that value of diameter which will deliver a nil value of error. In other words we are looking to change the value of diameter in cell B8 which will return a low value of error (the absolute value of the difference between given and computed pressure drop).

MS-Excel 'Goal Seek' facility offers a convenient procedure to iteratively obtain solution for such problems. That shall be presently demonstrated.

Select cell B14 and launch the Goal Seek facility which is found under 'Data', 'What-if Analysis'.

The dialog box provides you three boxes that need to be filled-in. In the 'Set cell' box enter B14 while the 'To value' cell should be provided by the value of the acceptable tolerance for your problem. We may enter the value 0.05 in this cell. In the 'By changing cell' enter cell address B8. Once we execute the Solver macro the value of pipe diameter converged to 1.162 m.

We may also try to use the 'Min' command by selecting that radio button and that will also lead to the same result.

Figure 3.3 shows the Goal Seek dialog box.

Discussion

A convenient facility was presented in the form of 'Goal Seek' that enables iterative solution to nonlinear engineering problems. You may use this Example as a base for bringing-in data from other engineering problems and also try to see the effect of changing the tolerance that you provide within the Goal seek facility.

Example 10.2 Interaction between a fan and duct carrying conditioned air— solution via optimization routine.

Problem definition

Figure 10.1 shows a schematic for a fan-duct system.
The static pressure, p_{static} (Pa) development in a given duct is given by Eq. (10.2.1),

$$p_{static} = 80 + 10.73Q^{1.8} \qquad (10.2.1)$$

A fan that has its characteristics described by Eq. 10.2.2 is to be matched with the above duct,

$$Q = 15 - 0.0000735\,p_{static}^2 \qquad (10.2.2)$$

where Q has the units of m³/s.
Solve the above system of nonlinear equations to obtain p_{static} and Q.

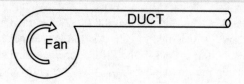

Fig. 10.1 A fan-duct system

Solution

Use the **Solver** facility that is available within MS-Excel to determine the operating point for the fan-duct system.

What we have here is a system of nonlinear equations that may quite easily be converted into an optimization problem. The steps are shown below:

Equations (10.2.1) and (10.2.2) may be manipulated to obtain Eqs. 10.2.1a and 10.2.1b,

$$\left[p_{static} - 80 - 10.73Q^{1.8}\right]^2 = 0 \qquad (10.2.1a)$$

$$\left[Q - 15 + 0.0000735\, p_{static}^2\right]^2 = 0 \qquad (10.2.2a)$$

We now introduce a function Z which is the sum of the left-hand sides (LHS) of the above two equations,

$$Z = \left[p_{static} - 80 - 10.73Q^{1.8}\right]^2 + \left[Q - 15 + 0.0000735\, p_{static}^2\right]^2$$

Z is the function that needs to be minimized, preferably to a nil value and those values of p_{static} and Q that return a minimum value of Z is the optimum solution.

Launch MS-Excel file Ex10-02.xls and explore the only sheet that is provided, i.e. 'fan-duct system'. Cells A2 through to E2 contain, respectively, the initial (assumed) values of p_{static} and Q, the LHS of Eqs. 10.1.1a and 10.1.1b, and the function Z.

The **Solver** facility is found under the **Data** menu. Once you have launched the **Solver** facility you need to specify the Objective Function via highlighting the cell E2 in the 'Set Objective' dialog box, selecting the 'Min' radio button, and then specifying cells A2:B2 in the 'By changing variable cells' dialog box,

You will note that 'Z' has the value of 0.00327 for the presently assumed values of p_{static} and Q, specified in cells A2 and B2, respectively.

Once the 'Solve' button is clicked the Z function is minimized to a value of 2.21E-10 and the optimized values of p_{static} and Q are now shown as 349.9 Pa and 5.999 m³/s.

Figure 10.2 shows the Solver dialog box.

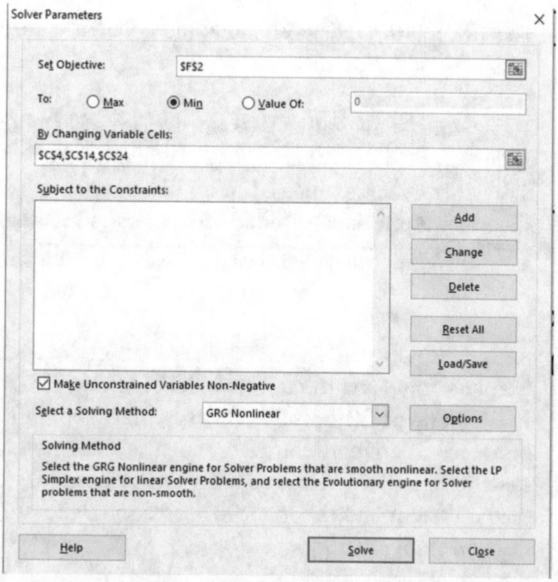

Fig. 10.2 Dialog box for Solver optimization tool

Discussion

You may now try the above procedure by providing various combinations of the starting values for p_{static} and Q, This will enable you to validate the robustness of the **Solver** facility.

Example 10.3 Interaction between a fan and duct carrying conditioned air— solution via VBA routine using successive substitution and fixed number of iterations.

Problem definition

Refer to the case of fan-duct system presented in Example 10.2. Solve the above system of nonlinear equations to obtain p_{static} and Q using the successive substitution technique.

Solution

The solution for this problem shall be attempted by using the successive substitution technique.

The file Ex10-03.xlsm contains the code for successive substitution routine presently under discussion.

The routine starts with an assumed value of Q which is used in Eq. 10.1.1 to obtain p_{static}. That value of p_{static} is then used to obtain Q via Eq. 10.1.2 and the whole cycle is repeated in an iterative loop. The convergence is checked such that the change in both parameters is less than, say, 0.1%.

Example 10.4 Flow through parallel pipes.

In fluid flow related problems, a system of pipes that are connected in parallel between a common entrance and discharge points is called a parallel-pipe system. In such systems, the pressure, or head loss, is the same for any branch of the system and the total flow or discharge is the cumulative sum of flow through individual pipes.

Problem definition

Let us consider three pipes that are connected in parallel between two points, A and B. The head loss between the two points is 6 m, i.e. the vertical distance between the pipe entry and exit is 6 m. The following data are provided for this system through which water flows at 20 °C. All units are in meter.

Length of pipe 1, $L_1 = 910$, diameter $D_1 = 0.3$, pipe roughness $\varepsilon_1 = 0.0003$.

Length of pipe 2, $L_2 = 610$, diameter $D_2 = 0.2$, pipe roughness $\varepsilon_2 = 0.00003$.

Length of pipe 3, $L_3 = 1220$, diameter $D_3 = 0.4$, pipe roughness $\varepsilon_3 = 0.00024$.

For a total flow of 340 L/s determine the flow through each of the three pipes.

Note that we may obtain the value of kinematic viscosity at 20 °C from any good thermo-physical property tables as 1.007×10^{-6} m^2/s.

Solution

Let us assume that the flow through pipe 1 is 85 L/s. The classical solution to such problems is the use the following iterative solution:

i. For pipe 1, using the assumed flow rate of 85 L/s calculate the fluid velocity, V_1 and relative roughness, ε_1/D_1.
ii. Obtain Reynolds number Re$_1$ and hence friction factor, f_1.
iii. Obtain head loss, $h_{f1} = [f_1 \, L_1 \, V^2_1] / (2g \, D_1)$.

iv. Use the above value of h_{f1} to obtain V_2 and V_3, then Re_2 and Re_3 for the remaining two pipes 2 and 3 after assuming their respective friction factors, f_2 and f_3, respectively.

v. Repeat step iv by re-calculating f_2 and f_3, then V_2, V_3, Re_1, and Re_2 until convergence is achieved.

vi. Using converged values of flow velocities obtain the respective discharges Q_1, Q_2, and Q_3.

With the kind of computational tools that are now available the above procedure may be considerably simplified via optimization techniques. The Excel-VBA file Ex10-04.xlsm presents solution to the current problem via use of the 'Solver' facility. 'Solver' uses an optimization technique to obtain solution for such problems where a number of independent variables are to be obtained once the 'Objective Function' is defined.

Open the file Ex10-04.xlsm which has four colored cells:

The blue-colored cells contain given data, while the yellow-colored cells contain assumed data, in this case the respective discharges Q_1, Q_2, and Q_3. Note that we seek the converged values of these discharges.

The green-colored cells contain the head loss for each of the three parallel pipes. In each of the three cases the value of h_f needs to be 6 m.

The two brown-colored cells, F1 and F2 contain the value of total flow rate, which is the sum of Q_1, Q_2, and Q_3, and the Objective Function which is defined herein as the sum of the absolute values of the difference between calculated value of h_f and given value of 6 for each of the three parallel pipes. We seek those values of Q_1, Q_2 and Q_3 that will return the minimum value of Objective Function. This is known as function minimization. For clarity of presentation the Swamee-Jain formula for friction is also included within the Excel sheet.

Select cell F2, then select 'Data' from the top bar of the menu, then select 'Solver'. Note that if Solver tool is not visible then you will have to install it using a simple procedure that is shown on the web. A screen dump of the present Solver dialog box is shown in Fig. 10.2. The Set Objective contains cell F2 where our Objective Function is placed, while By Changing Variable Cells contain reference to the three cells for providing discharge quantities for the three parallel pipe. Note that function minimization has been chosen by selecting the 'Min' radio button.

The last thing to note is that there are three optimization techniques available:

- GRG Nonlinear
- Simplex LP
- Evolutionary

Select 'GRG Nonlinear; which is most appropriate for this type of problem dealing with nonlinear minimization.

Once you click the 'Solve' button you will note that the Objective Function has attained a nil value and the solution is Q_1, = 99.4, Q_2 = 54.2, and Q_3 = 187.2 L/s. The total of the computed flows is 340.8 which is within 0.2% of the given flow which was 340 L/s.

10.1 Conclusion

Fluid mechanics is an established science that finds applications in civil, mechanical, chemical, and nuclear engineering practice. Starting from simple problem solving such as obtaining friction factor for flow within any given pipe or duct, this chapter presented advanced applications such as designing for enhancing the flow through branched pipes and obtaining the requirements of pump power for any given duty.

Exercises

E10.1 Refer to Example 10.2. Go through the solution presented therein.

Zafar Grand is a ventilation engineer employed by Malakpet Corporation. Zafar needs to match the characteristics of a fan and duct system.

The component equations are given below for duct (Eq. E10.1.1) and fan (Eq. E10.1.2) are:

$$P = 0.0625 + 0.653Q^{1.8} \tag{E10.1.1}$$

$$P = 0.3 - 0.2Q^2 \tag{E10.1.2}$$

Find the values of Q and P that satisfy the above equations.

Hint: You may combine the above two equations in a single nonlinear equation which may then be solved either by bisection method or via use of MS-Excel Goal Seek facility.

Answer: $Q = 0.5$ m^3/s, $P = 0.25$ kPa.